Advanced Issues
in Partial Least Squares
Structural Equation
Modeling

Sara Miller McCune founded SAGE Publishing in 1965 to support the dissemination of usable knowledge and educate a global community. SAGE publishes more than 1000 journals and over 800 new books each year, spanning a wide range of subject areas. Our growing selection of library products includes archives, data, case studies and video. SAGE remains majority owned by our founder and after her lifetime will become owned by a charitable trust that secures the company's continued independence.

Los Angeles | London | New Delhi | Singapore | Washington DC | Melbourne

Advanced Issues in Partial Least Squares Structural Equation Modeling

Joseph F. Hair, Jr.
University of South Alabama, USA

Marko Sarstedt
Otto-von-Guericke-University Magdeburg, Germany, and University of Newcastle, Australia

Christian M. Ringle
Hamburg University of Technology, Germany, and University of Newcastle, Australia

Siegfried P. Gudergan
University of Newcastle, Australia

Los Angeles | London | New Delhi
Singapore | Washington DC | Melbourne

FOR INFORMATION:

SAGE Publications, Inc.
2455 Teller Road
Thousand Oaks, California 91320
E-mail: order@sagepub.com

SAGE Publications Ltd.
1 Oliver's Yard
55 City Road
London EC1Y 1SP
United Kingdom

SAGE Publications India Pvt. Ltd.
B 1/I 1 Mohan Cooperative Industrial Area
Mathura Road, New Delhi 110 044
India

SAGE Publications Asia-Pacific Pte. Ltd.
3 Church Street
#10-04 Samsung Hub
Singapore 049483

Acquisitions Editor: Leah Fargotstein
Editorial Assistant: Yvonne McDuffee
eLearning Editor: Laura Kirkhuff
Production Editor: Kelly DeRosa
Copy Editor: Sarah J. Duffy
Typesetter: C&M Digitals (P) Ltd.
Proofreader: Jennifer Grubba
Indexer: Will Ragsdale
Cover Designer: Anupama Krishnan
Marketing Manager: Susannah Goldes

Printed in the United States of America

Library of Congress Cataloging-in-Publication Data

Names: Hair, Joseph F., author.

Title: Advanced issues in partial least squares structural equation modeling / Joseph F. Hair, Jr., University of South Alabama, [and three others].

Description: Los Angeles : SAGE, [2018] | Includes bibliographical references and index.

Identifiers: LCCN 2016050545 | ISBN 9781483377391 (pbk. : alk. paper)

Subjects: LCSH: Least squares. | Structural equation modeling.

Classification: LCC QA275 .H235 2018 | DDC 511/.42—dc23
LC record available at https://lccn.loc.gov/2016050545

This book is printed on acid-free paper.

17 18 19 20 21 10 9 8 7 6 5 4 3 2 1

Brief Contents

Detailed Contents

Preface

Researchers in the fields of management, marketing, management information systems, and other social science disciplines have been increasingly focused on seeking to better understand the complex interrelationships that constitute the "black box" of a variety of organizational or behavioral aspects. Partial least squares structural equation modeling (PLS-SEM) has enabled researchers to simultaneously estimate such complex interrelationships involving a variety of constructs and indicators with their direct, indirect, or moderating relationships that would otherwise not be easy to disentangle and examine. However, recent research focuses on more fully understanding and also explaining the roles of intervening and contingent variables and relationships among variables that are of interest to researchers. For example, greater interest has been placed on unraveling the contingencies that are reflected in differences that characterize subgroups of individuals, organizations, or environments. To understand such contingencies requires confidently assessing observed or unobserved heterogeneity. In a similar vein, a common conceptualization recognizes that effects are not necessarily constant but that they might diminish or increase such that researchers need to move beyond linear modeling to nonlinear modeling.

This emergence of more complex modeling requirements goes hand in hand with and underlines the critical importance of advanced analytical methods. Notable advances in PLS-SEM include, for example,

- confirmatory tetrad analysis to empirically assess the mode of measurement (Gudergan, Ringle, Wende, & Will, 2008),

- the heterotrait-monotrait ratio of correlations as a new approach to test discriminant validity (Henseler, Ringle, & Sarstedt, 2015),

- prediction-oriented segmentation analysis to identify and treat unobserved heterogeneity (Becker, Rai, Ringle, & Völckner, 2013),

- different types of multigroup analysis (Sarstedt, Henseler, & Ringle, 2011), and

- invariance testing by means of the measurement invariance of composite models approach (Henseler, Ringle, & Sarstedt, 2016).

Many of these advanced analysis techniques such as treatment of unobserved heterogeneity, multigroup comparisons, invariance, and confirmatory tetrad analysis have been available in covariance-based SEM (CB-SEM), but only recently have methodologists started introducing them in a PLS-SEM context. At the same time, other approaches such as the importance-performance map analysis and continuous moderator analyses are exclusive to PLS-SEM.

The benefits of having such advanced PLS-SEM approaches readily at hand are tremendous, since these types of analyses assist in the evaluation of PLS-SEM estimations and are increasingly being requested by editors and reviewers. At the same time, however, applying these and other advanced PLS-SEM approaches requires understanding their intricacies and knowing when they can assist in analyzing data in a meaningful way such that the applications fit the research context.

Along with the development of advanced techniques, research has recently witnessed an increasing debate about the relative advantages of PLS-SEM vis-á-vis other SEM methods. Such scientific debates are important as they serve as a catalyst that sparks further careful examination of the method's properties. While oftentimes a better understanding of the advantages and disadvantages emerges, additional research and methodological advances also stem from objective and constructive discussions among scholars, which aim at moving science forward. Recently, however, the scholarly community has witnessed a surprising level of uninformed discussion concerning PLS-SEM. Antonakis, Bendahan, Jacquart, and Lalive (2010, p. 1103) allude that "there is no use for PLS whatsoever [and] thus strongly encourage researchers to abandon it." Other authors similarly suggest that the use of PLS-SEM "is very difficult to justify" (Rönkkö & Evermann, 2013, p. 443). Leaving aside the tone of these and similar statements, which aim at shutting down any discussion, thereby revealing a rather disturbing lack of understanding of science, they show that critics of PLS-SEM frequently offer incorrect and unfounded rationale for avoiding the method (Rigdon, 2016). These misconceptions have their roots in a lack of understanding of the method's conceptual underpinnings and particularly the measurement philosophy it relies on (e.g., Rigdon, 2012; Sarstedt, Hair, Ringle, Thiele, & Gudergan, 2016).

"The PLS and ML-LISREL approaches to path models with latent variables indirectly observed by multiple indicators are complementary rather than competitive. The key difference of PLS relative to LISREL is the explicit estimation of the case values of the latent variables" (Wold, 1982, p. 5). More precisely, PLS is a composite-based approach to SEM in that it linearly combines indicators to form composite variables (Lohmöller, 1989), which serve as proxies for the concepts under investigation (Rigdon, 2016). This approach is different from CB-SEM, which is common factor–based and therefore considers the constructs as common factors that explain the covariation between their associated indicators. While this distinction has long been noted (e.g., Jöreskog & Wold, 1982; Schneeweiß, 1991), only recently has research started clarifying the implications of this differentiation for PLS-SEM use (Rigdon, 2012; Sarstedt, Hair, Ringle, Thiele, & Gudergan, 2016; Hair, Hult, Ringle, Sarstedt, & Thiele, in press). At the same time, recent research has brought forward different approaches to adjust PLS-SEM estimates to be the same as CB-SEM when estimating common factor models (e.g., Bentler & Huang 2014; Dijkstra & Henseler 2015a, 2015b; Dijkstra & Schermelleh-Engel, 2014).

We wrote this advanced book on PLS-SEM as an extension of the *Primer on Partial Least Squares Structural Equation Modeling (PLS-SEM)*, 2nd edition (Hair, Hult, Ringle, & Sarstedt, 2017), for several reasons. First, ways of using PLS-SEM have advanced significantly in recent years. But as these methodological advances have been developing so quickly, guidance on applying these emerging approaches appropriately is just not available, particularly in an easily readable, user-friendly format. This book is meant to equip researchers with a sound understanding of and the competencies to apply advanced PLS-SEM approaches. Second, as research questions are becoming more sophisticated and data more readily available, researchers need to apply more advanced SEM analyses, which are very often possible only with PLS-SEM. This book not only introduces and clarifies methodological advances but importantly also outlines when such advanced PLS-SEM approaches are and are not applicable. Third, there is a lot of misunderstanding about when PLS-SEM approaches are more suitable than are other SEM methods and vice versa. Such misapprehension can result in SEM approaches being incorrectly applied and interpreted. As a consequence, researchers may lack confidence in employing PLS-SEM in general and advanced modeling approaches associated with PLS-SEM as well.

To summarize, this book enables researchers to develop the confidence to apply advanced PLS-SEM approaches when appropriate and to do so correctly. It clarifies those research contexts for which PLS-SEM is not necessarily the appropriate SEM approach, but also the contexts for which PLS-SEM is not only the absolutely correct method, but indeed the only suitable SEM approach currently available. Ultimately, the book enables researchers to write research reports and journal articles that correctly apply advanced PLS-SEM approaches, when appropriate, and communicate their findings more effectively. Throughout the book we stress that any advanced PLS-SEM approach should not be applied blindly, but with great care in that it should fit the appropriate research context and the data characteristics that underpin the research.

The approach of this book follows the *Primer on Partial Least Squares Structural Equation Modeling (PLS-SEM)*, 2nd edition (Hair et al., 2017), and is based on our many years of conducting and teaching research as well as our desire to communicate both the foundations and advances of the PLS-SEM method to a much broader audience. To accomplish this goal, we have limited the emphasis on equations, formulas, Greek symbols, and so forth that are typical of most books and articles on this topic. Instead, we explain in detail the fundamentals of advanced PLS-SEM analyses and provide rules of thumb that can be used as general guidelines for understanding and evaluating the results of applying the method.

As a further effort to facilitate learning, we use a single case study throughout the book. The case is drawn from a published study on corporate reputation (Eberl, 2010) and we believe is sufficiently general to be understood by researchers coming from many different areas of social science and other fields, thus further facilitating comprehension of the method. All examples in this book use a single software package, SmartPLS 3 (www.smartpls.com), that can be used not only to complete the exercises in this book but also to execute the reader's own research. Review and critical thinking questions are posed at the end of the chapters, and key terms are defined to help the reader develop a better understanding of the concepts. Finally, suggested readings and extensive references are provided so that the interested researcher can appreciate a more advanced coverage of the topic.

The book chapters and learning support supplements are organized around the learning outcomes shown at the beginning of each chapter. Moreover, instead of a single summary at the end of each chapter, we present a separate topical summary for each learning outcome. This approach makes the book more understandable and usable for both students and teachers. The website for the book (www.pls-sem.com) also includes other support materials to facilitate learning and applying the PLS-SEM method.

We would like to acknowledge the many insights and suggestions provided by a number of our colleagues and students. Most notably, we thank Sönke Albers (Kühne Logistics University), Haya Ajjan (Elon University), Murad Ali (King Abdulaziz University), Necmi Avkiran (University of Queensland), Barry J. Babin (Louisiana Tech University), Jan-Michael Becker (University of Cologne), Diógenes de Souza Bido (Universidade Presbiteriana Mackenzie), Roger Calantone (Michigan State University), Gabriel Cepeda Carrión (University of Seville), Wynne W. Chin (University of Houston), Alain Yee Loong Chong (University of Nottingham), Adamantios Diamantopoulos (University of Vienna), Theo Dijkstra (University of Groningen), Markus Eberl (TNS Infratest), Andreas Eggert (University of Paderborn), Vincenzo Esposito Vinzi (ESSEC Business School), David Garson (North Carolina State University), Oliver Götz (Reutlingen University), Anne Gottfried (University of Southern Mississippi), Lars Grønholdt (Copenhagen Business School), Karl-Werner Hansmann (University of Hamburg), Sven Hauff (University of Hamburg), Jörg Henseler (University of Twente), Roland Holten (Goethe University Frankfurt), Lucas Hopkins (Florida State University), Geoffrey S. Hubona (R-Courseware), G. Tomas M. Hult (Michigan State University), Ida Rosnita Ismail (Universiti Kebangsaan Malaysia), Dave Ketchen (Auburn University), Hengky Latan (Diponegoro University), Marcel Lichters (Otto-von-Guericke-University Magdeburg), Anne Martensen (Copenhagen Business School), Annette Marie Mills (University of Canterbury), Arthur Money (Henley Business School), Erik Mooi (University of Melbourne), Christian Nitzl (Universität der Bundeswehr München), Arun Rai (Georgia State University), Sascha Raithel (Freie Universität Berlin), Thurasamy Ramayah (Universiti Sains Malaysia), S. Mostafa Rasoolimanesh (Universiti Sains Malaysia), Nicole F. Richter

(University of Southern Denmark), Edward E. Rigdon (Georgia State University), James A. Robins (Vienna University of Economics and Business), José Luis Roldán (University of Seville), Alexander Rossmann (Reutlingen University), Phillip Samouel (University of Kingston), Gastón Sánchez (University of California, Berkeley), Rainer Schlittgen (University of Hamburg), Tobias Schütz (ESB Business School), Manfred Schwaiger (Ludwig-Maximilians-Universität München), Harjit Sekhon (Coventry University), Abdel Monim Shaltoni (Alfaisal University), Mac (Wen-Lung) Shiau (Ming Chuan University), Galit Shmueli (National Tsing Hua University), Rudolf R. Sinkovics (Alliance Manchester Business School), Donna Smith (Ryerson University), Detmar W. Straub (Georgia State University), Michel Tenenhaus (HEC Paris), Nelly Todorova (University of Canterbury), Laura Trinchera (NEOMA Business School), Sven Wende (SmartPLS GmbH), and Anita Whiting (Clayton State University) for their helpful remarks. In addition, we thank the following reviewers for their very useful comments: David Hemsworth (Nipissing University), Rajesh Iyer (Bradley University), Jay Memmott (University of South Dakota), Brent Smith (Saint Joseph's University), and James N. Smith (Nova Southeastern University). Also, we thank the team of doctoral student and research fellows at Hamburg University of Technology (TUHH) and Otto-von-Guericke-University Magdeburg—namely, Kati Barth, Katrin Engelke, Andreas Fischer, Amra Kramo-Čaluk, Frauke Kühn, Anke Lepthien, Doreen Neubert, Jana Rosenbusch, Victor Schliwa, Sandra Schubring, and Kai Oliver Thiele—for their kind support. Finally, at SAGE we thank our entire publishing team, most notably Kelly DeRosa, Sarah J. Duffy, Leah Fargotstein, and Yvonne McDuffee.

We hope this book will expand knowledge of the capabilities and benefits of PLS-SEM to a much broader group of researchers and practitioners. Last, if you have any remarks, suggestions, or ideas to improve this book, please get in touch with us. We appreciate any feedback on the book's concept and contents!

Joseph F. Hair Jr., *University of South Alabama*

Marko Sarstedt, *Otto-von-Guericke-University Magdeburg, Germany, and University of Newcastle, Australia*

Christian M. Ringle, *Hamburg University of Technology, Germany, and University of Newcastle, Australia*

Siegfried P. Gudergan
University of Newcastle, Australia

About the Authors

Joseph F. Hair, Jr. is professor of marketing, DBA director, and the Cleverdon Chair of Business in the Mitchell College of Business, University of South Alabama. He previously was senior scholar, DBA Program, in the Coles College of Business, Kennesaw State University, and held the Copeland Endowed Chair of Entrepreneurship and was director of the Entrepreneurship Institute in the Ourso College of Business Administration, Louisiana State University. He has authored over 60 books, including *Multivariate Data Analysis* (7th ed., 2010), *MKTG* (10th ed., 2016), *Essentials of Business Research Methods* (2016), and *Essentials of Marketing Research* (4th ed., 2017). He also has published numerous articles in scholarly journals and was recognized as the Academy of Marketing Science marketing educator of the year. A popular guest speaker, Professor Hair often presents seminars on research techniques, multivariate data analysis, and marketing issues for organizations in Europe, Australia, China, India, and South America.

Marko Sarstedt is chaired professor of marketing at the Otto-von-Guericke-University Magdeburg (Germany) and conjoint professor at the Faculty of Business and Law at the University of Newcastle (Australia). He previously was an assistant professor of quantitative methods in marketing and management at the Ludwig-Maximilians-University Munich (Germany). His main research is in the application and advancement of structural equation modeling methods to further the understanding of consumer behavior and to improve marketing decision making. His research has been published in journals such as the *Journal of Marketing Research*, *Journal of the Academy of Marketing Science*, *Organizational Research Methods*, *MIS Quarterly*, *International Journal of Research in Marketing*, *Long Range Planning*, *Journal of World Business*, and *Journal of Business Research*. According to the Handelsblatt ranking, Professor Sarstedt is among the top three young academic marketing researchers in Germany, Austria, and Switzerland. He regularly teaches doctoral seminars on multivariate statistics, structural equation modeling, and measurement worldwide.

Christian M. Ringle is a chaired professor of management at the Hamburg University of Technology (Germany) and a conjoint professor at the Faculty of Business and Law at the University of Newcastle (Australia). He holds a master's degree in business administration from the University of Kansas and received a PhD from the University of Hamburg (Germany). His widely published research addresses the management of organizations, strategic and human resource management, marketing, and quantitative methods for business and market research. He is a cofounder of SmartPLS (www.smartpls.com), a software tool with graphical user interface for the application of the partial least squares structural equation modeling (PLS-SEM) method. Besides supporting consultancies and international corporations, he regularly teaches doctoral seminars on multivariate statistics, the PLS-SEM method, and the use of SmartPLS. More information about Professor Ringle and his full list of publications can be found at www.tuhh.de/hrmo/team/prof-dr-c-m-ringle.html

Siegfried P. Gudergan is a chaired professor of strategy at Newcastle Business School in the University of Newcastle (Australia). He previously was head of school and professor of marketing at UTS Business School in the University of Technology, Sydney (Australia). He holds a PhD from the Australian Graduate School of Management that was awarded by both the University of Sydney and the University of New South Wales. His research focuses on managerial decision making and is applied to strategic management issues. Some of his work that concerns PLS-SEM has appeared in journals such as the *Journal of the Academy of Marketing Science*, *Long Range Planning*, and *Journal of Business Research*. He has been the primary advisor for doctoral students who have been awarded, for example, an honorable mention as finalist for the Dissertation Award of the Academy of Management's Business Policy & Strategy Division and the Best PhD Paper Award by the Strategic Management Society.

Visit the author-created companion website to download data sets and other materials to supplement your studies in PLS-SEM! study.sagepub.com/hairadvanced

CHAPTER 1

An Overview of Recent and Emerging Developments in PLS-SEM

LEARNING OUTCOMES

1. Understand the origins and evolution of PLS-SEM.

2. Comprehend the principles of and recent developments in measurement.

3. Get to know the essential differences between PLS-SEM and CB-SEM and understand when to use each method.

CHAPTER PREVIEW

In recent years many developments have occurred in partial least squares structural equation modeling (PLS-SEM). Among the most prominent was the latest release of the SmartPLS software (Ringle, Wende, & Becker, 2015). User-friendly software makes social science scholars much more efficient and effective in reaching their research and publication goals. This is especially true when the software includes options to complete the most up-to-date analyses possible with a particular statistical method.

In this chapter, we first provide an overview of the origins and evolution of PLS-SEM to establish a foundation for better understanding why the method was slow to be adopted but has been increasingly applied in recent years across many social science disciplines, particularly the various fields of business administration. We then summarize the software that facilitates easy application of this rapidly emerging technique and briefly highlight recent methodological developments in PLS-SEM. To understand how PLS-SEM differs from covariance-based SEM (CB-SEM), we then discuss different approaches to measure conceptual variables and highlight which method is more suitable for certain model types.

ORIGINS AND EVOLUTION OF PARTIAL LEAST SQUARES STRUCTURAL EQUATION MODELING

The precursors to the PLS-SEM method were two iterative procedures that used least squares estimation to develop solutions for single and multicomponent models and for canonical correlation (Wold, 1966). Further development of these procedures by Herman Wold led to the nonlinear iterative partial least squares (NIPALS) algorithm (Wold, 1973). A revised generalized version of the **PLS algorithm** focused on establishing and including latent variables in path models (Lohmöller, 1989; Wold, 1980, 1982, 1985).

Several PLS procedures evolved from Wold's generalized least squares algorithm (Mateos-Aparicio, 2011). One procedure is **principal component regression,** which performs a principal component analysis on the independent variables and in which the principal components are used as predictive/explanatory variables for the dependent variable. But principal component regression focuses on reducing the dimensionality of the independent variables without taking into account the relationship between the independent and dependent variables.

Another procedure is **partial least squares regression (PLS-R),** which was originally designed to reduce the problem of multicollinearity in regression models. PLS-R is an approach that focuses on dimension reduction of the independent variables in a regression model, with the objective of removing multicollinearity from the predictor variables, but doing so in a manner that optimizes the variance extracted from the independent variables while simultaneously maximizing the variance explained in the dependent variables. More precisely, PLS-R relies on a principal component analysis that extracts linear composites of the independent variables and their respective scores, reduces the dimensionality of the independent variables, and takes into consideration the relationship between the independent and dependent variables, thus maximizing the explanation of the variance in the dependent variable. As a result, PLS-R allows researchers to estimate models with many more independent variables than observations in the data set (Valencia & Diaz-Llanos, 2003).

Interestingly, PLS-R was not developed by Herman Wold but by his son, Svante Wold, in the early 1980s. Svante Wold was working in the field of analytical chemistry that is known today as chemometrics—the application of statistical methods to chemical data. Together with Harald Martens, he adapted NIPALS to analyze chemical data, and in addition to solving the problem of multicollinearity in multiple regression models, their method solved the problem that arises when the number of variables is larger than the number of respondents.

A third procedure that emerged from Wold's generalized PLS algorithm was **partial least squares path modeling**, which later became known as **PLS-SEM** (Hair, Ringle, & Sarstedt, 2011). PLS-SEM determines the parameters of a set of equations in a path model by combining principal component analysis to assess the measurement models with path analysis to estimate the relationships between latent variables. Wold (1982) proposed his "soft model basic design" underlying PLS-SEM as an alternative to Jöreskog's (1973) **covariance-based SEM (CB-SEM)**. CB-SEM is sometimes referred to as **factor-based SEM,** and has been labeled as hard modeling because of its much more stringent assumptions in terms of data distribution and sample size. Importantly, "it is not the concepts nor the models nor the estimation techniques which are 'soft,' only the distributional assumptions" (Lohmöller, 1989, p. 64). While both approaches were developed about the same time, CB-SEM became much more widely applied because of its early availability through the LISREL software since the late 1970s. In contrast, the first software for PLS-SEM was LVPLS, which appeared in the mid-1980s (Lohmöller, 1984) but was not very user-friendly. It was not until Chin's (1994) PLS-Graph software in the mid-1990s that PLS-SEM began being more widely applied. With the release of SmartPLS 2 in 2005 (Ringle, Wende, & Will, 2005), PLS-SEM applications grew exponentially.

Exhibit 1.1 summarizes the application of PLS-SEM in the top journals in marketing (Hair, Sarstedt, Ringle, & Mena, 2012) and strategic management (Hair, Sarstedt, Pieper, & Ringle, 2012), as well as *MIS Quarterly*, the flagship journal in management information systems research (Ringle, Sarstedt, & Straub, 2012). However, PLS-SEM use extends to many other fields as diverse as supply chain management, operations management, and family business research (e.g., Do Valle & Assaker, 2016; Hair, Hollingsworth, Randolph, & Chong, 2017; Henseler, Ringle, & Sarstedt, 2012; Kaufmann & Gaeckler, 2015; Lee, Petter, Fayard, & Robinson, 2011; Nitzl, 2016; Peng & Lai, 2012; Richter, Sinkovics, Ringle, & Schlägel, 2016; Sarstedt, Ringle, Smith, Reams, & Hair, 2014). Also, several journals such as *Journal of Marketing Theory and Practice* (Hair et al., 2011), *Long Range Planning* (Hair, Ringle, & Sarstedt, 2013; Robins, 2012, 2014), *Journal of Business Research* (Carrión, Henseler, Ringle, &

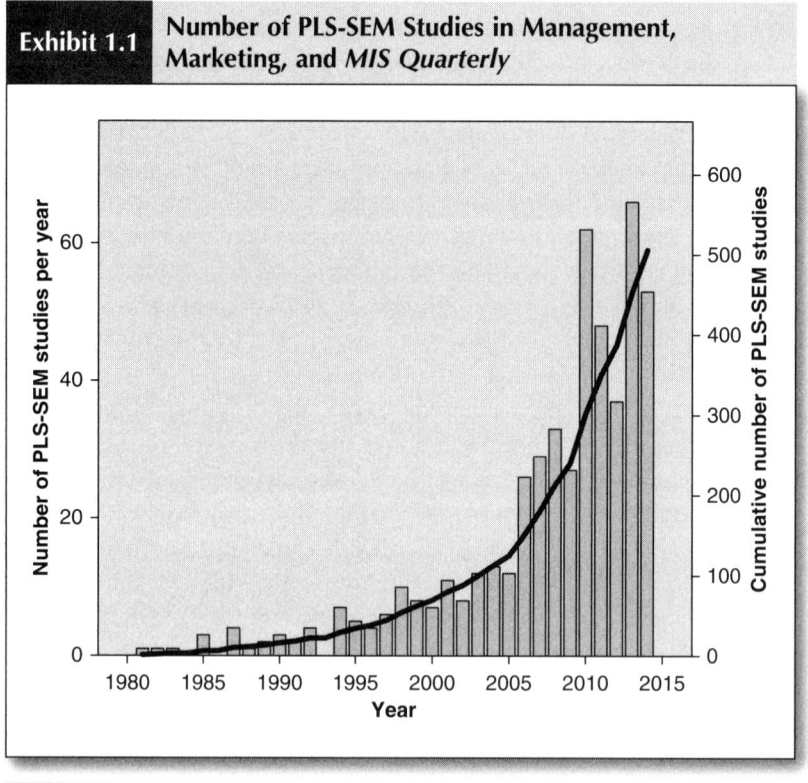

Exhibit 1.1   **Number of PLS-SEM Studies in Management, Marketing, and *MIS Quarterly***

Note: PLS-SEM studies published in *MIS Quarterly* were only considered from 1992 on.

Roldán, 2016), and *International Marketing Review* (Richter et al., 2016) published special issues on the PLS-SEM method. Besides its popularity in business research, the use of PLS-SEM, as published journal articles reveal, has recently expanded into other research areas such as biology (e.g., Kansky, Kidd, & Knight, 2016), engineering (e.g., Kwofie, Adinyira, & Fugar, 2015; Liu, Zhao, & Yan, 2016; Minkman, Rutten, & van der Sanden, 2016), medicine (e.g., Hsu, Chang, & Lai, 2016; Pai, 2016; Pedrosa, Rodrigues, Oliveira, & Alexandre, 2016), and psychology (e.g., Jisha, & Thomas, 2016; Willaby, Costa, Burns, MacCann, & Roberts, 2015).

The release of version 3 of the most widely applied PLS-SEM software, SmartPLS, includes many new features. First, obtaining a basic solution from version 2 (Ringle et al., 2005) of the software was straightforward, and continues to be true for version 3. But quite a few diagnostics for assessing PLS-SEM solutions when using version 2 had to be calculated manually or completed with another software package, such as Excel, SPSS, or Statistica. SmartPLS 3 automatically performs these functions, including f^2 effect size, assessment of multi-collinearity, and so forth. More important, however, is the addition of options such as confirmatory tetrad analysis (Gudergan, Ringle, Wende, & Will, 2008), the new heterotrait-monotrait ratio of correlations (HTMT) criterion to test discriminant validity (Henseler, Ringle, & Sarstedt, 2015), prediction-oriented segmentation (Becker, Rai, Ringle, & Völckner, 2013), moderator analysis (Henseler & Chin, 2010), different types of multigroup analysis (Sarstedt, Henseler, & Ringle, 2011), and invariance testing by means of the measurement invariance of composite models approach (Henseler, Ringle, & Sarstedt, 2016). The benefits of having these options readily at hand are tremendous, since these types of analyses are increasingly being requested by journal editors and reviewers. These developments are very important for social science researchers because many of them are not possible with CB-SEM due to its limiting assumptions, while the others have not been programmed to facilitate easy use.

MEASUREMENT

Conceptual Variables and Proxies

Whether researchers follow a deductive or inductive research approach, at some point—in their search to better understand and explain theory—they deal with theoretical models and conceptual variables. A **theoretical model** reflects a set of structural relationships, usually based on a set of equations connecting conceptual variables that formalize a theory and visually represent the relationships (Bollen, 2002). As elements of theoretical models, **conceptual variables** represent broad ideas or thoughts about abstract concepts that researchers establish and propose to measure in their research (e.g., customer satisfaction).

Constructs represent conceptual variables in statistical models such as in a structural equation model. They are intended to enable empirical testing of hypotheses regarding the conceptual variables (Rigdon, 2012) and are conceptually defined in terms of attribute and object (e.g., MacKenzie, Podsakoff, & Podsakoff, 2011). The **attribute** defines the general type of property to which the concept refers, such as an attitude (e.g., attitude toward an advertisement), a perception (e.g., perceived ease of use of technology), or a behavioral intention (e.g., purchase intention). The **focal object** is an entity to which the property is applied. For example, the focus of interest could be a customer's satisfaction with the products, satisfaction with the services, and satisfaction with the prices. In these examples, satisfaction constitutes the attribute, whereas products, services, and prices represent the focal objects.

Establishing a construct definition also includes determination of the dimensionality that describes the conceptual variable, with each dimension representing a different aspect of the conceptual variable. For example, the concept commitment can be viewed as encompassing three different dimensions (Meyer & Allen, 1991), each defined and measured differently: affective commitment (i.e., emotional attachment), continuance commitment (i.e., the gains versus losses of working at an organization), and normative commitment (i.e., based on feelings of obligation). Each of these dimensions requires a separate operational definition. A conceptual variable is not per se characterized as unidimensional or multidimensional, let alone two-, three-, or four-dimensional (Bollen, 2011). Rather it depends on the context-specific definition of the conceptual variable and the denotation that comes with it. The denotation can, in principle, be infinite since the same conceptual variable may represent different levels of theoretical abstraction across contexts (Diamantopoulos, 2005; Law & Wong, 1999). An operational definition is subject to the context within which a conceptual variable is examined, and the definition can change from one study to another. Consequently, an operational definition can differ in terms of dimensionality and the object of interest, depending on the context of the study. For example, a customer's satisfaction with the service can be broken down into more concrete subdimensions, such as satisfaction with the speed of service, the servicescape, and the staff. The latter dimension can be divided into more concrete subdimensions such as satisfaction with the friendliness, competence, and outer appearance of the service staff. Each of these aspects can, in principle, be further broken down into yet more concrete subdimensions.

As the construct definition clarifies how the abstract, conceptual variable relates to measurable, observable quantities, it also assists in understanding how, in structural equation models, conceptual variables are represented by **constructs** (also referred to as **latent variables**). Constructs are not directly observed but rather inferred (mathematically) from **manifest variables** that are observed (directly measured). Manifest variables are also referred to as **items** or **indicators**. The indicators correspond, for example, to questions in a survey capturing respondents' perceptions, attitudes, and behaviors. Importantly, manifest variables do not need to be survey-based data but can also be secondary data (e.g., satisfaction ratings from a review website). When a conceptual variable is multidimensional, each dimension needs to be represented by a separate construct, operationalized with a specific set of manifest variables. For example, if a study considers all three dimensions of commitment, then there will be three separate latent variables in the structural equation model.

The construct definition explains how the conceptual variable should relate to manifest variables from a theoretical perspective. However, constructs do not represent conceptual variables perfectly as any concept and any operational definition has some degree of ambiguity associated with it. Furthermore, constructs stem from data and therefore share the data's idiosyncrasies (Cliff, 1983; MacCallum, Browne, & Cai, 2007; Rigdon, 2012), which further detaches them from the concepts they intend to represent. In this context, Michell (2013, p. 20) notes that constructs are contrived in a way that is detached from the actual structure of testing phenomena and held in place by an array of quantitative methods, such as factor analysis, which gratuitously presume quantitative structure rather than infer it from the relevant phenomena.

Against this background, Rigdon (2012, pp. 343–344) concludes that constructs should rather be viewed as "something created from the empirical data which is intended to enable empirical testing of propositions regarding the concept." That is, all measures of conceptual variables are approximations of or proxies for conceptual variables, independent from how they were derived (e.g., Wickens, 1972). Thus, irrespective of the quality with which a conceptual variable is theoretically substantiated and operationally defined and the rigor that encompasses measurement model development, any measurement in structural equation models produces only proxies for latent variables (Rigdon, 2012). This assessment is in line with the proliferation of all sorts of instruments that claim to measure essentially the same construct, albeit often with little chance to convert one instrument's measures into any other instrument's measures (Salzberger, Sarstedt, & Diamantopoulos, 2016). For example, business research and practice has brought forward a multitude of measurement instruments for corporate reputation, which rest on the same definition of the concept but differ fundamentally in terms of their underlying conceptualizations and measurement items (Sarstedt, Wilczynski, & Melewar, 2013).

Measurement Models

Following the construct definition, the next step is to conceptualize a **measurement model**, which expresses how to measure the construct by means of a set of indicators. Generally, there are two broad ways to conceptualize measurement models. The first approach is referred to as reflective measurement. In a reflective measurement model the indicators are considered to be error-prone manifestations of an underlying construct with relationships going from the construct to its indicators (Exhibit 1.2). The relationship between an observed and an unobserved variable is usually modeled as expressed in the following equation:

$$x = l \cdot Y + e,$$

where x is the observed indicator variable, Y is the latent variable, the loading l is a regression coefficient quantifying the strength of the relationship between each indicator x and its construct Y; e represents the random measurement error.

Exhibit 1.2 illustrates a **reflective measurement model** for a latent variable Y_1, measured with four indicators x_1, x_2, x_3 and x_4. Here, the l_1 to l_4 relationships result from four simple regressions, as depicted by the previous equation, with each indicator x_1 to x_4 as dependent variables and construct Y_1 as independent variable. **Reflective indicators** (sometimes referred to as **effect indicators** in the psychometric literature) can be viewed as a representative sample of all the possible items available in the conceptual domain of the construct (Nunnally & Bernstein, 1994). Since a reflective measurement model dictates that all items reflect the same construct, indicators associated with a particular construct should be highly correlated with each other. In addition, individual items should be interchangeable, and any single item can generally be left out without changing the meaning of the construct, as long as the construct has sufficient reliability. The fact that the relationship goes from the construct to its indicators implies that if the evaluation of the latent trait changes (e.g., because of a change in the standard of comparison), all indicators will change simultaneously—at least to some extent.

The other type of measurement model is formative measurement. In a **formative measurement model** the indicators form the

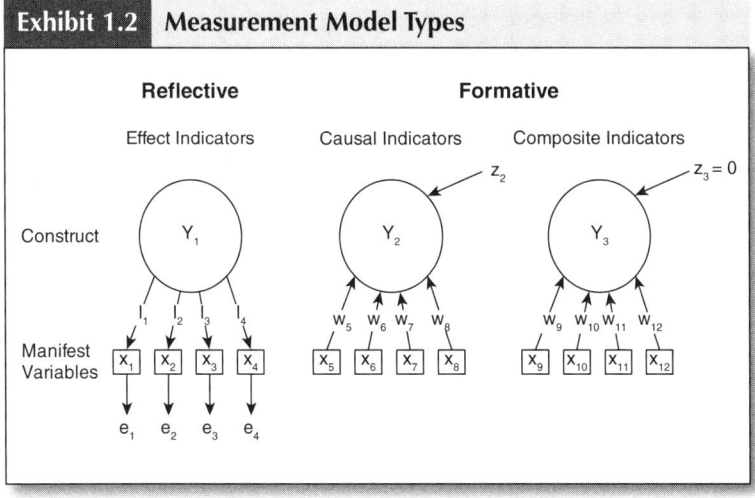

Exhibit 1.2 **Measurement Model Types**

Reflective — Effect Indicators

Formative — Causal Indicators, Composite Indicators

Construct: Y_1, Y_2, Y_3

Manifest Variables

construct by means of linear combinations. A change in an indicator's value due to, for example, a change in a respondent's assessment of the trait being captured by the indicator changes the value of the construct. That is, variation in the indicators precedes variation in the latent variable (Borsboom, Mellenbergh, & van Heerden, 2003), which means that, by definition, constructs with a formative measurement model are inextricably tied to their measures (Diamantopoulos, 2006). Besides the difference in the relationship between indicator(s) and construct, formative measurement models do not require correlated indicators.

Despite these clear conceptual differences, deciding whether to specify measurement models reflectively or formatively is not clear-cut in practice, as conceptual variables do not inherently follow a reflective or formative measurement logic. Rather, the researcher has the flexibility to define how such proxies are to be derived in a measurement model based on the construct definition the researcher specifies. Consider, for example, the concept of perceived switching costs. Jones, Mothersbaugh, and Beatty (2000, p. 262) define perceived switching costs as "consumer perceptions of the time, money, and effort associated with changing service providers." Their measurement approach in the context of banking services draws on three items, which constitute reflections or consequences of perceived switching costs ("In general it would be a hassle changing banks," "It would take a lot of time and effort changing banks," and "For me, the costs in time, money, and effort to switch banks are high"). Hence, the authors implicitly assume that there is a concept of perceived switching costs, which can be manifested by querying a set of (e.g., three) items. Barroso and Picón (2012, p. 532), on the other hand, consider perceived switching costs as "a latent aggregate construct that is expressed as an algebraic composition of its different dimensions." These authors identify a set of six dimensions (benefit loss costs, personal relationship loss costs, economic risks costs, evaluation costs, setup costs, and monetary loss costs), which represent certain specific characteristics, each covering an independent part of the perceived switching costs concept. As such, Barroso and Picón's conceptualization of perceived switching costs represents a formative measurement model logic. Of course, the underlying items are empirically correlated, and perhaps causally related, but they are not actually exchangeable in the way the reflective measurement model conceptualization assumes they are (Rigdon, Preacher, et al., 2011). That is, their correlation is not because the construct of perceived switching costs is assumed to be their common cause.

Further contributing to the difficulties of deciding on the measurement perspective is the fact that there is not one type of formative measurement model—as suggested in the early works on formative measurement (e.g., Diamantopoulos & Winklhofer, 2001) and the use of formative measurement models in statistical analysis (e.g., Hair et al., 2011). Rather, two types of indicators exist in formative measurement models: causal indicators and composite indicators (Bollen, 2011; Bollen & Bauldry, 2011). As implied by their name, **causal indicators** are assumed to cause the underlying construct. Therefore, the indicators should have conceptual unity in that all the indicators correspond to the researcher's definition of the concept (Bollen & Diamantopoulos, 2016). Breadth of coverage of the domain is extremely important to ensure that the domain of content is adequately captured—omitting important indicators implies omitting a part of the conceptual variable that the construct represents. As causal indicators are expected to cover all aspects of the content domain (Bollen & Bauldry, 2011), constructs measured with causal indicators (construct Y_2 with indicators x_5 to x_8 in Exhibit 1.2) have an error term (z_2 in Exhibit 1.2). This error term captures all the other causes of the construct not included in the model (Diamantopoulos, 2006). Importantly, the indicators themselves are assumed to be error-free. The following equation represents a measurement model with causal indicators, where w_i indicates the contribution of x_i ($i = 1, \ldots, I$) to Y, and z is an error term associated with Y:

$$Y = \sum_{i=1}^{I} w_i \bullet x_i + z.$$

The other indicator type, **composite indicators**, closely resembles causal indicators except for a small detail. Similar to constructs measured with causal indicators, constructs measured with composite indicators (construct Y_3 with indicators x_9 to x_{12} in Exhibit 1.2) also have an error term but this term is set to zero ($z_3 = 0$ in Exhibit 1.2). This distinction has an essential implication for the characterization of formative measurement models (Henseler et al., 2014), since composite indicators operate as contributors to a construct rather than truly causing it (Bollen, 2011; Bollen & Bauldry, 2011). They form the composite representing the construct fully by means of linear combinations, thereby ensuring that the construct has no error term. The following equation illustrates a measurement model with composite indicators, where Y is a linear combination of indicators x_i, each with an indicator weight w_i (Rigdon, 2012):

$$Y = \sum_{i=1}^{I} w_i \cdot x_i.$$

Traditionally, composite indicators have been viewed as a means to combine several variables to represent some new entity whose meaning is defined by the choice of indicators (e.g., Bollen, 2011; Henseler, Hubona, & Ray, 2016). For example, a measurement model conceptualization of information search activities could be based on capturing the sum of the activities that customers engage in when seeking information from dealers, promotional materials, the Internet, and other sources. Another researcher might choose a different set of variables to form a measure of information search activities. Thus, the items ultimately determine the meaning of the construct, which implies that adding or omitting an indicator potentially alters the nature of the construct. Therefore, according to Bollen (2011), composite indicators need not share conceptual unity but may have a similar theme.

In practice, however, it remains largely unclear where to draw a line between items having conceptual unity and sharing a similar theme. In fact, researchers naturally adopt a construct definition of the concept and choose items in operationalizing measurement models that match their definition, regardless of whether the actual measurement conceptualization draws on reflective, causal, or composite indicators. That is, they treat the constructs in their studies as unitary entities, just like Barroso and Picón (2012) do when offering an in-depth literature review of the nature and dimensionality of the perceived switching costs concept, prior to deriving indicators in their operationalization of the construct's measurement model. As such, they follow best practice in measurement by developing a guiding conceptual framework and set of indicators, which imbues the construct's theoretical meaning.

Assuming that researchers use measures of composite indicators merely as convenient summaries of the data—as some researchers do—implies that the common practice of aggregating items as composites to represent constructs, even though commonly done in practically all non-SEM studies in all fields of research (e.g., when using composites as input for a regression analysis), is without any theoretical justification and undermines the fundamentals of appropriate measurement. However, the very same measures have mostly been carefully developed and tested following common measurement model evaluation guidelines—as extensively documented in standard measurement scale handbooks (e.g., Bearden, Netemeyer, & Haws, 2011; Bruner, James, & Hensel, 2001). Thus, the very activity of forming composites from validated measurement scales interweaves composite and causal indicators. Therefore, calling for an abandonment of PLS-SEM because it uses composites as representations for conceptual variables (Guide & Ketokivi, 2015) implies an abandonment of practically all empirical research involving latent variables, including studies published in operations management, marketing, strategy, and other social sciences.

Composite indicators not only offer a convenient way to summarize the data but can also be used to measure any type of property to which the focal concept refers, including attitudes, perceptions, and behavioral intentions. However, as with any type of measurement conceptualization, researchers need to offer a clear definition and define items that closely match this definition—that is, they must share conceptual unity. Alternatively, measurement models with composite indicators can be interpreted as a prescription for dimension reduction, where the aim is to condense the measures so they adequately cover a conceptual variable's salient features (Dijkstra & Henseler, 2011). For example, a researcher may be interested in measuring the salient aspects of perceived switching costs by means of three (composite) indicators, which cover aspects particularly relevant to the study at hand (e.g., evaluation costs, setup costs, monetary loss costs).

Exhibit 1.3 extends the previous display of reflective and formative measurement model conceptualizations to include the conceptual variables they seek to represent (Rigdon, 2012). Conceptual variables are the small circles at the top of the diagrams. The proxy estimates are shown as the smaller eight-point star-like symbols in the middle of the larger circles. The vertical double-headed arrows suggest the extent to which the proxies are valid measures—the shorter these vertical arrows, the more valid the proxies.

The position of the proxy within the larger circle is an indication of the reliability of the construct measures, and the middle represents

Exhibit 1.3	Measurement Model Types and Conceptual Variables

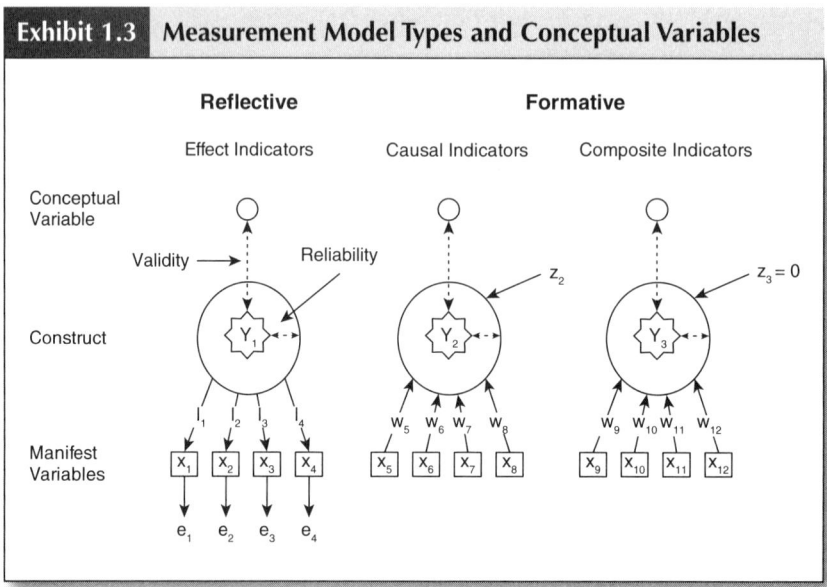

perfect reliability. Note that we have to consider different types of reliability, depending on the model type. Internal consistency reliability statistics such as Cronbach's alpha apply to reflective measurement models. But these statistics are not appropriate when evaluating formative measurement models because causal and composite indicators do not necessarily have to correlate. In formative measurement models, test-retest reliability is the only means of testing the measures' reliability. However, in light of the manifold problems of assessing test-retest reliability (e.g., Campbell & Stanley, 1966), most researchers routinely disregard reliability assessment when estimating formative measurement models.

MODEL ESTIMATION

The previous sections describe different measurement model types and their indicators. The next question is how to estimate these measurement models and their relationships with others in the structural equation model. Researchers typically use two approaches to estimate structural equation models: CB-SEM (Bollen, 1989; Diamantopoulos, 1994; Jöreskog, 1978) and PLS-SEM (Hair, Hult, Ringle, & Sarstedt, 2017; Lohmöller, 1989; Wold, 1982).

While both methods estimate the relationships among constructs and indicators, they differ fundamentally in their approaches in doing so (Jöreskog & Wold, 1982).

CB-SEM initially divides the variance of each indicator into two parts: (1) the **common variance**, which is estimated from the variance shared with other indicators in the measurement model of a construct, and (2) the **unique variance**, which consists of both **specific variance** and **error variance**. The specific variance is assumed to be systematic and reliable, while the error variance is assumed to be random and unreliable (i.e., measurement, sampling, and specification error). CB-SEM initially calculates the covariances of a set of variables (common variance), and only that variance is included in any solutions derived. CB-SEM, therefore, follows a **common factor model** approach in the estimation of the construct measures, which hypothesizes that the variance of a set of indicators can be perfectly explained by the existence of one unobserved variable (the common factor) and individual random error (Spearman, 1927; Thurstone, 1947). The common factor model estimation approach fully conforms to the measurement philosophy underlying reflective measurement models.

In principle, CB-SEM can also accommodate formative measurement models with causal indicators, but to ensure model identification, researchers must follow model specification rules that require certain constraints on the model (Bollen & Davis, 2009; Diamantopoulos & Riefler, 2011). MacCallum and Browne (1993), among others, have addressed these CB-SEM specification issues in detail. However, as Hair, Sarstedt, Ringle, and Mena (2012, p. 420) note, "[t]hese constraints often contradict theoretical considerations, and the question arises whether model design should guide theory or vice versa." As an alternative, CB-SEM scholars have proposed the **multiple indicators and multiple causes (MIMIC) model,** which includes both formative and reflective indicators (e.g., Diamantopoulos & Riefler, 2011; Diamantopoulos, Riefler, & Roth, 2008). While MIMIC models enable researchers to deal with the identification problem, they do not overcome the problem that formative measurement models with causal indicators invariably underrepresent the variance in the construct, since correlated indicators are required by the CB-SEM common factor model to produce a valid proxy and thereby to adequately represent a conceptual variable. As Lee and Cadogan (2013, p. 243) note, "researchers should not be misled into thinking that achieving statistical identification allows one to obtain information about the variance of a formative latent variable." In summary, application of the MIMIC approach to include constructs measured with causal indicators in CB-SEM invariably introduces a bias in the proxies and therefore in the SEM solutions. Clearly, at best, CB-SEM only allows for approximating formative measurement models with causal indicators.

Similarly, CB-SEM can accommodate formative measurement models with composite indicators (Bollen & Diamantopoulos, 2017). Since constructs measured with composite indicators are defined by having zero variances, the identification of the construct's error variance is not an issue. However, problems arise with regard to the identification of all paths leading to as well as flowing out from the construct. Grace and Bollen (2007, p. 206) note that this problem can be solved by specifying a single incoming or outgoing path relationship to 1.0 (e.g., one relationship from the indicators to the composite per measurement model). While such specifications overcome parameter identification issues, they severely limit the interpretability of the structural model estimates with regard to the significance and magnitude of the fixed paths. Furthermore, constructs with composite indicators can only be included as exogenous latent variables while at the same time requiring at least one relationship to a reflectively measured endogenous latent variable. Both of these issues limit their usefulness in CB-SEM applications.

Different from CB-SEM, PLS-SEM does not divide the variance into common and unique variance. More precisely, the objective of PLS-SEM is to account for the total variance in the observed indicators rather than to explain only the correlations between the indicators (e.g., Chin, 1998; Tenenhaus, Esposito Vinzi, Chatelin, & Lauro, 2005). The differing treatment of the variances in PLS-SEM compared to CB-SEM corresponds to the distinction between principal component analysis and common factor analysis. Whereas principal components analysis assesses total variance and estimates factors (components) as linear combinations of indicators, common factor analysis assesses shared (common) variance only (Kline, 2015). The logic of the PLS-SEM approach is that in estimating the model relationships, all of the variance (common, unique, and error) that the exogenous variables have in common with the endogenous variables should be included. The underlying notion of PLS-SEM is that the indicators can be (linearly) combined to form composite variables that are comprehensive representations of the latent variables, and that these linear combinations are valid proxies of the conceptual variables under investigation. Therefore, PLS-SEM follows a **composite model approach** in the estimation of the construct measures. PLS-SEM's designation as composite-based only refers to the method's way to represent constructs that approximate the conceptual variables in a model. But while PLS-SEM draws on composites whose use has traditionally been considered to be consistent with formative measurement models and not reflective measurement models, the fact is the method readily accommodates both measurement model types (e.g., Roldán & Sánchez-Franco, 2012). In formative measurement models, however, PLS-SEM treats all indicators as composite indicators. That is, the method does not allow for the explicit modeling of a construct's error term measured with causal indicators (i.e., the error term z_3 in Exhibits 1.2 and 1.3 is zero). As a consequence and analogous to CB-SEM, PLS-SEM only allows for approximating formative measurement models with causal indicators.

Finally, to estimate the model parameters, PLS-SEM uses two modes, which relate to the way the method estimates the indicator weights that represent each indicator's contribution to the composite. Mode A corresponds to correlation weights derived from bivariate correlations between each indicator and the construct; Mode B corresponds to regression weights, the standard in ordinary least squares regression analysis. Regression weights take into account not only the correlation between each indicator and the construct but also the correlations between the indicators. No matter which mode for estimating the indicator weights is used, the resulting latent variable is always modeled as a composite (Henseler, Hubona, & Rai, 2016). That is, since all multi-item measures are converted into weighted components—even in Mode A—PLS-SEM computes components by means of linear combinations of indicators.

PLS-SEM by default uses Mode A for reflectively measured constructs and Mode B for formatively measured constructs. Recent research, however, suggests that selecting the appropriate weighting mode requires a more thoughtful approach. Specifically, Becker, Rai, and Rigdon (2013) show that for formatively measured constructs, Mode A estimation yields better out-of-sample prediction for sample sizes larger than 100 and when the R^2 value is moderate to large (i.e., $R^2 > 0.30$). For large sample sizes and large R^2 values, Mode A and Mode B perform equally well in terms of out-of-sample prediction. In terms of parameter accuracy in the structural model, Mode A performs best when sample size or R^2 values are small to medium. For larger sample sizes or R^2 values, Mode A and Mode B estimations do not differ in terms of parameter accuracy.

The option of using different estimation modes per construct allows tailoring the model estimation to fit the study's objective. Similarly, item weights of certain construct measures can also be defined a priori (Howell, Breivik, & Wilcox, 2013), thereby maintaining temporal constancy in the relative importance of each item in the formation of the composite. While such a step facilitates a consistent interpretation in repeated assessments of the same model, fixed weights equalize any differences in each item's importance for forming the composite in a specific context. However, understanding individual item weights offers important insights in the specific context of the analysis. When measuring, for example, service satisfaction, weights denote which individual items are of particular importance in the shaping of service satisfaction. Although individual item weights likely differ across contexts, as all items share conceptual unity, the empirical meaning of the construct (e.g., Howell, Breivik, & Wilcox, 2007) will differ only within the realm of the construct definition.

PLS-SEM OR CB-SEM?

Researchers have long acknowledged the differences between PLS-SEM and CB-SEM, highlighting situations that favor the use of one method over the other (e.g., Jöreskog & Wold, 1982). Similarly, an abundance of studies have empirically compared the differences in model estimates produced by CB-SEM versus PLS-SEM. These comparisons univocally focused on PLS-SEM's capabilities to obtain the same results as CB-SEM. However, PLS-SEM in its original form (Wold, 1982; Lohmöller, 1989) was not created to mimic CB-SEM but rather follows a different philosophy of measurement and aim of the analysis. In the following sections, we discuss these two key criteria that primarily distinguish PLS-SEM from CB-SEM. Hair et al. (2017) offer a more detailed discussion of model and measurement characteristics relevant for choosing between PLS-SEM and CB-SEM. Finally, Rigdon (2016) and Sarstedt, Ringle, and

Hair (2017) discuss reasons for using PLS-SEM and flawed arguments in favor of the method.

Philosophy of Measurement

A crucial conceptual difference between PLS-SEM and CB-SEM relates to the way each method treats the latent variables included in the model. As CB-SEM considers only the indicators' common variance, the method treats the constructs as common factors. PLS on the other hand uses the indicators' total variance in that the method generates linear combinations of indicators to represent the constructs, thereby constituting a composite model approach to SEM.

Several simulation studies have shown that PLS-SEM measurement model estimates (loadings) are larger and structural model estimates (path coefficients) are smaller relative to those obtained by CB-SEM (e.g., Reinartz, Haenlein, & Henseler, 2009). However, these simulation studies univocally defined common factor populations and drew data from these populations (Marcoulides & Chin 2013). Therefore, they evaluated PLS-SEM on the basis of (common factor–based) populations that are inconsistent with the method's philosophy of measurement. Indeed, practically every statistical method will perform less well when the underlying model is misspecified, and PLS-SEM is no exception in this regard. In fact, CB-SEM is known to not perform well when the model being estimated is inconsistent with the population, such as when the data are generated from a composite model (Becker, Rai, & Rigdon, 2013; Henseler et al., 2014). Becker, Rai, and Rigdon (2013) defined composite-based populations, and showed that PLS-SEM parameter estimates are the same as those of CB-SEM—converging on those values as sample size increases. Hair, Hult, Ringle, Sarstedt, and Thiele (2017) provide support for these findings in their comparative evaluation of composite-based SEM methods. Thus, the term *consistency at large* when applied to PLS-SEM results as representing bias is a misnomer.

CB-SEM results based on the common factor model, including the path coefficients, inter-construct correlations, and indicator loadings, have been labeled as the "true score" (Bollen, 1989). But what makes them so—other than that they have been called that previously? What is a true score when measuring concepts such as attitudes, perceptions, or intentions? Researchers have long warned that the common factor model rarely holds in applied research (Schönemann & Wang, 1972).

For example, among 72 articles published during 2012 in what Atinc, Simmering, and Kroll (2012) consider the four leading management journals that tested one or more common factor model(s), less than 10% contained a common factor model that did not have to be rejected. In light of these and similar results, neither model can be assumed to carry greater significance than the other with regard to the existence or nature of conceptual variables (Rigdon, 2016). Any construct operationalization—regardless of whether based on a common factor or composite models logic—comes with an abundance of ambiguities related to, for example, the construct definition (e.g., Diamantopoulos, 2005; Gilliam & Voss 2013), the item wordings, and the number of items necessary to capture the construct domain (e.g., DeVellis, 2011). The same holds for the measurement validation, which is highly context-specific, thereby capitalizing on the idiosyncrasies of the data at hand. In light of these ambiguities, it is more reasonable to view estimates produced by both CB-SEM and PLS-SEM as proxies for the concepts under research and nothing more (Rigdon, 2012).

Irrespective of these conceptual concerns, the differences that PLS-SEM produces when estimating common factor models is very small provided that the measurement models meet minimum recommended standards in terms of the number of indicators and indicator loadings, and the model is correctly specified. Specifically, when the measurement models have four or more indicators and indicator loadings meet recommended standards, there is very little bias in parameter estimates when estimating a common factor model with PLS-SEM, as shown, for example, by Reinartz et al. (2009).

Prior efforts to dramatize the differences between CB-SEM and PLS-SEM estimates (Rönkkö, McIntosh, Antonakis, & Edwards, 2016) in Reinartz et al.'s (2009) study focused on descriptive differences between population values and parameter estimates only, disregarding the simple concept of statistical inference. As Reinartz et al. (2009, p. 318) note in their results description of all simulation conditions, "parameter estimates do not differ significantly from their theoretical values for either ML-based CB-SEM (p-values between 0.3963 and 0.5621) or PLS-SEM (p-values between 0.1906 and 0.3449)". Only when the model estimation draws on a high sample size ($N = 10,000$) and measurement models with many indicators with high loadings, statistically significant differences between 0.58% and 3.23% occurred. Most importantly, when estimating common factor models, the divergence between PLS-SEM and CB-SEM results is of little practical relevance for the vast majority of applications (e.g., Astrachan, Patel, & Wanzenried, 2014). In fact, recent research shows that the bias that CB-SEM produces when (incorrectly) using the method to estimate composite models is much higher than the bias that PLS-SEM produces when using the method to (incorrectly) estimate common factor models (Sarstedt, Hair, Ringle, Thiele, & Gudergan 2016). The same research shows that PLS-SEM's parameter bias is clearly lower when correctly using the method to estimate composite models than CB-SEM's bias when using the method for estimating common factor models, especially when sample sizes are small. To summarize, PLS-SEM introduces practically no bias when estimating data from a composite model population, regardless of whether the measurement models draw on composite indicators or effect (i.e., reflective) indicators. Biases are somewhat higher for factor model populations, but low in absolute terms. Clearly, PLS-SEM is optimal for estimating composite models while simultaneously allowing approximating common factor models with effect indicators with practically no limitation (Sarstedt, Hair, et al., 2016).

The minimal magnitude of differences between composite-based PLS-SEM and common factor–based CB-SEM when the underlying data stems from a common factor model population calls into question the need for "corrections" of the PLS-SEM method when estimating factor models. Specifically, in an effort to align common factor and composite-based SEM methods, Dijkstra and Henseler (2015a, 2015b) recently introduced the **consistent PLS** (**PLSc**) approach (also see Bentler & Huang, 2014). PLSc follows a composite modeling logic but mimics a common factor model. To do so, the method first estimates the model parameters using the standard PLS-SEM algorithm and corrects these estimates for attenuation using the consistent reliability coefficient ρ_A. This correction applies only to reflective measurement models; formative measurement model estimates of composite indicators remain unchanged. That is, when estimating reflective measurement models, PLSc draws on the questionable premise that CB-SEM results are the benchmark against which PLS-SEM results should be evaluated. Several researchers have noted that PLS-SEM-based indicator loadings are upwardly biased, but that is in comparison to indicator loadings from CB-SEM. The divergence of PLS-SEM parameter estimates should not be considered a bias, but a different result based on an algorithm following a different philosophy of measurement.

Generally, it is difficult to argue why PLS-SEM users would want the method to produce the same results as CB-SEM, since that method is already widely recognized and accepted as an approach to obtain solutions to research designed around the common factor model. One potential reason might be that the model is underidentified or its estimation leads to convergence problems in CB-SEM. Alternatively, researchers may want to include common factors and composites in the same model and ensure that each type is estimated in accordance with its conceptualization.

Aim of the Analysis

Apart from the differences in the philosophy of measurement and the different treatment of latent variables, the nature of **latent variable scores** as produced by PLS-SEM and CB-SEM also has consequences for the methods' areas of application. In CB-SEM latent variable scores are not unique, which means that there is an infinite number of different sets of latent variable scores that will fit the model equally well. A crucial consequence of this **factor (score) indeterminacy** is that the correlations between a common factor and any variable outside the factor model are indeterminate (Guttman, 1955). That is, a correlation may be high or low, depending on which set of factor scores one chooses. As a result, this limitation makes CB-SEM extremely unsuitable for prediction (e.g., Dijkstra, 2014). In contrast, PLS-SEM always produces a single specific (i.e., determinate) score for each composite for each observation, once the weights are established. Using these proxies as input to obtain a solution, PLS-SEM applies ordinary least squares regression with the objective of maximizing the R^2 values of the endogenous constructs. PLS-SEM is therefore the preferred method when the aim of the analysis is prediction (Albers, 2010; Rigdon, 2014; Sarstedt, Ringle, Henseler, & Hair, 2014).

Researchers should be aware that the R^2 values in both types of SEM cannot be interpreted the same as with multiple regression models. The R^2 value in SEM is the variance explained in the construct scores, and not the variance explained in the indicators. Since the construct score represents only a portion of the variance in the indicators, the R^2 value in SEM is the amount of variance explained in the construct. The amount of variance included in the construct score is represented by the **average variance extracted** (AVE), which is the degree to which the construct explains the variance of its indicators. Thus the R^2 value in SEM is the variance explained in the AVE.

As an example of the R^2 value in SEM, consider the following. CB-SEM is based only on common variance. If you assume common variance representing the indicators of an endogenous construct is 60% of the total variance, and the AVE is 0.60, then the variance in the AVE is actually 36% of the original variance in the total variance (60% of 60% = 36%). Thus, if the CB-SEM R^2 value is moderately high, for example, 0.50, then the proportion of the total variance predicted in the endogenous construct is only 18% of the total variance in the indicators. In contrast, consider a similar example with PLS-SEM that is based on the total variance. If the AVE of an endogenous construct is 0.60, and the PLS-SEM R^2 value is 0.50, then the proportion of the total variance predicted in the endogenous construct is 30% of the total variance in the indicators of the endogenous construct, almost twice as much. PLS-SEM is based on total variance, and AVEs are always higher than in CB-SEM, so the R^2 value in PLS-SEM is more meaningful as a predictor of the variance included in the indicators of the endogenous constructs than is the R^2 value in CB-SEM. To summarize, when the research objective is prediction, the choice of methods is clear—it should be PLS-SEM.

ORGANIZATION OF THE REMAINING CHAPTERS

The remaining chapters provide more detailed information on advanced analyses using PLS-SEM, including specific examples of how to use the SmartPLS 3 software for such analyses. In this advanced book, we expand on Stage 7 of the systematic procedure for applying PLS-SEM (Exhibit 1.4) from the *Primer on Partial Least Squares Structural Equation Modeling,* 2nd edition *(PLS-SEM)* (Hair et al., 2017). The advanced PLS-SEM analyses will enable you to better understand and explain your results, and provide the types of analysis and diagnostic metrics editors and reviewers increasingly request. Exhibit 1.5 identifies the chapters and topics covered in this book.

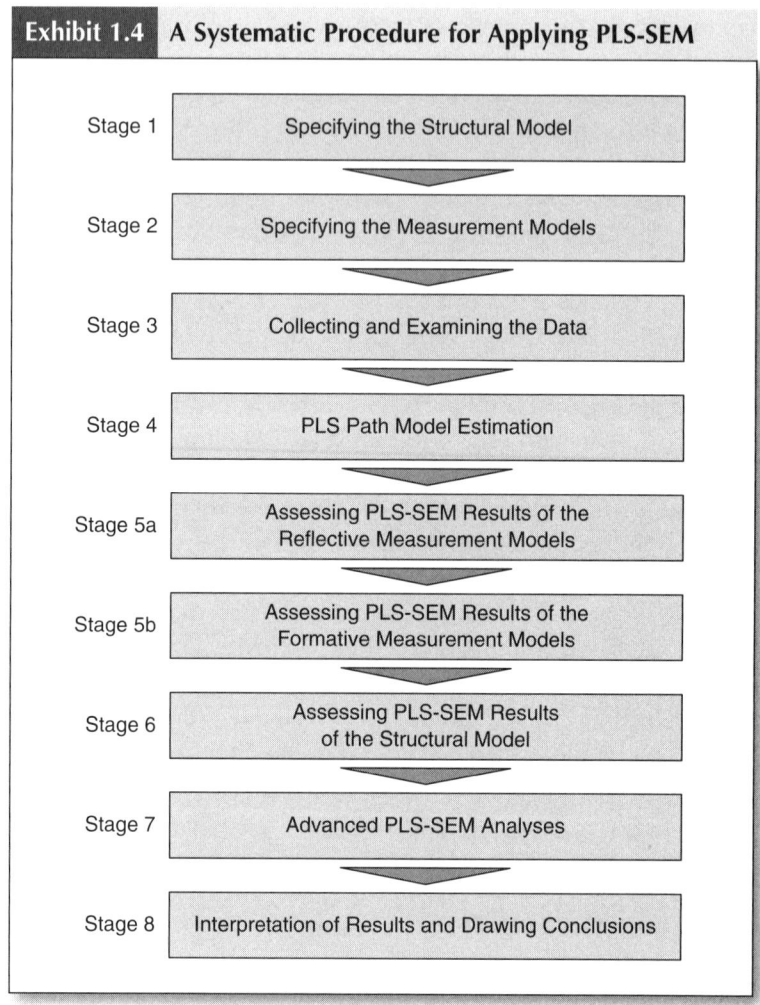

Exhibit 1.4   A Systematic Procedure for Applying PLS-SEM

Stage 1 — Specifying the Structural Model

Stage 2 — Specifying the Measurement Models

Stage 3 — Collecting and Examining the Data

Stage 4 — PLS Path Model Estimation

Stage 5a — Assessing PLS-SEM Results of the Reflective Measurement Models

Stage 5b — Assessing PLS-SEM Results of the Formative Measurement Models

Stage 6 — Assessing PLS-SEM Results of the Structural Model

Stage 7 — Advanced PLS-SEM Analyses

Stage 8 — Interpretation of Results and Drawing Conclusions

Chapter 2 offers an introduction to advanced modeling topics, starting with hierarchical component models. A hierarchical component model represents a more general construct, measured at a higher level of abstraction, while simultaneously including several subcomponents, which cover more concrete traits of the conceptual variable represented by this construct. With the increasing complexity of theories and cause-effect models in the social sciences, researchers have more often used these models in their PLS-SEM studies (e.g., Ringle et al., 2012). Similarly, there has also been increased interest in modeling nonlinear relationships between constructs. When the relationship

Exhibit 1.5	Chapters and Topics in This Book
Chapter	*Topics*
2	Advanced Modeling – Hierarchical component models – Nonlinear relationships
3	Advanced Model Assessment – Confirmatory tetrad analysis (CTA-PLS) – Importance-performance map analysis (IPMA)
4	Modeling Observed Heterogeneity – Measurement invariance assessment (MICOM) – Multigroup analysis
5	Modeling Unobserved Heterogeneity – Finite mixture PLS (FIMIX-PLS) – PLS prediction-oriented segmentation (PLS-POS)

between two constructs is nonlinear, the size of the effect between two constructs depends not only on the magnitude of change in the exogenous construct but also on its value. In the second part of Chapter 2, we introduce the principles of nonlinear modeling and describe how to run corresponding analyses in SmartPLS 3.

In Chapter 3, we discuss two types of advanced model assessment, starting with the confirmatory tetrad analysis (CTA-PLS), which allows empirically assessing whether the data support a formative measurement model specification or a reflective specification. Next, we introduce the importance-performance map analysis (IPMA), which extends the standard structural model results reporting of path coefficient estimates by adding a dimension that considers the average values of the latent variable scores.

The following two chapters deal with different concepts that enable researchers to model heterogeneous data. Chapter 4 first provides an overview of observed and unobserved heterogeneity, showing how disregarding heterogeneous data structures can provoke biased results. Next, we discuss measurement invariance, which is a primary concern before comparing groups of data. The chapter concludes with an introduction of different types of multigroup analysis that are used to compare parameters (usually path coefficients) between two or more groups of data. While these methods allow accounting for observed heterogeneity, more often than not, situations arise in which differences related to unobserved heterogeneity prevent the derivation of accurate results as the analysis on the aggregate data level masks group-specific effects. Chapter 5 introduces two methods, finite mixture PLS (FIMIX-PLS) and prediction-oriented segmentation in PLS (PLS-POS), that enable researchers to identify and treat unobserved heterogeneity in PLS path models.

CASE STUDY ILLUSTRATION

Corporate Reputation Model

The most effective way to learn how to use a statistical method is to apply it to a set of data. Throughout this book, we use a single example that enables you to do that. The example is drawn from a series of published studies on corporate reputation, which is general enough to be understood by researchers from many different disciplines, thus further facilitating comprehension of the analyses presented in this book. More precisely, we draw on the corporate reputation model by Eberl (2010), which Hair et al. (2017) use in their primer on PLS-SEM. The model's purpose is to explain the effects of corporate reputation on customer satisfaction (*CUSA*) and, ultimately, customer loyalty (*CUSL*). Corporate reputation represents a company's overall evaluation by its stakeholders (Helm, Eggert, & Garnefeld, 2010). Following Schwaiger (2004), corporate reputation is measured using two dimensions. One dimension represents the cognitive evaluations of the company and measures the construct describing the company's competence (*COMP*). The second dimension captures affective judgments and assesses perceptions of the company's likeability (*LIKE*). Schwaiger further identifies four antecedent dimensions of reputation: quality (*QUAL*), performance (*PERF*), attractiveness (*ATTR*), and corporate social responsibility (*CSOR*). Exhibit 1.6 shows the corporate reputation model.

The measurement models of the *LIKE*, *COMP*, and *CUSL* constructs have three reflective indicators, whereas *CUSA* is measured with a single item. Note that you should generally avoid using single items, particularly in PLS-SEM analyses (e.g., Diamantopoulos, Sarstedt, Fuchs, Kaiser, & Wilczynski, 2012; Sarstedt, Diamantopoulos, & Salzberger, 2016; Sarstedt, Diamantopoulos, Salzberger, & Baumgartner, 2016). However, we included this single item for illustrative purposes. Finally, the four exogenous constructs are measured by a total of 21 formative indicators. Exhibit 1.7 provides an overview of all items and item wordings. Respondents rated the questions on 7-point Likert

Exhibit 1.6   Corporate Reputation Model

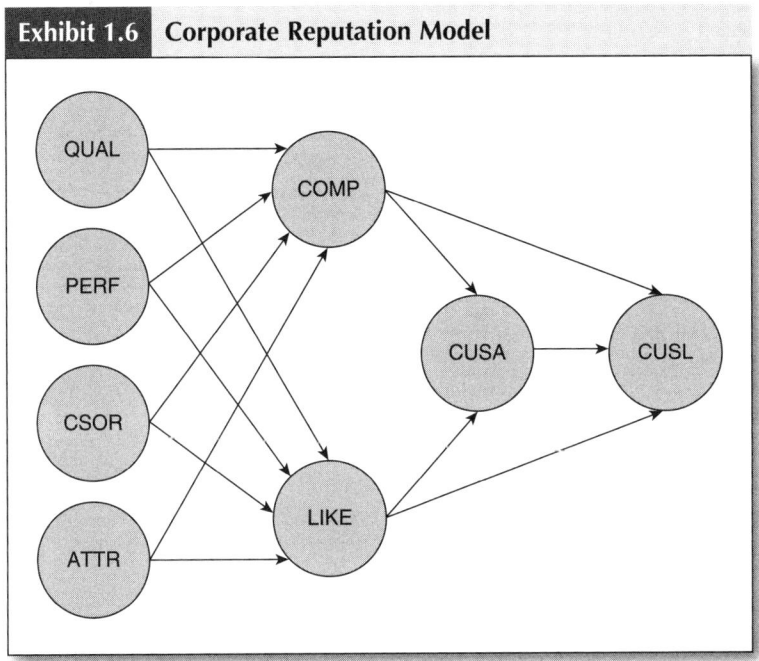

scales, with higher scores denoting higher levels of agreement with a particular statement. In the case of the *cusa* indicator, higher scores denote higher levels of satisfaction. Satisfaction and loyalty were measured with respect to the respondents' own service providers.

Exhibit 1.7   Items and Item Wordings

Competence *(COMP)*	
comp_1	[The company] is a top competitor in its market.
comp_2	As far as I know, [the company] is recognized worldwide.
comp_3	I believe that [the company] performs at a premium level.
Likeability (*LIKE*)	
like_1	[The company] is a company that I can better identify with than other companies.
like_2	[The company] is a company that I would regret more not having if it no longer existed than I would other companies.
like_3	I regard [the company] as a likeable company.

(Continued)

Exhibit 1.7	(Continued)
Customer Loyalty (CUSL)	
cusl_1	I would recommend [the company] to friends and relatives.
cusl_2	If I had to choose again, I would choose [the company] as my mobile phone services provider.
cusl_3	I will remain a customer of [the company] in the future.
Customer Satisfaction (CUSA)	
cusa	If you consider your experiences with [the company], how satisfied are you with [the company]?
Quality (QUAL)	
qual_1	The products/services offered by [the company] are of high quality.
qual_2	[The company] is an innovator, rather than an imitator with respect to [industry].
qual_3	[The company]'s products/services offer good value for money.
qual_4	The services [the company] offered are good.
qual_5	Customer concerns are held in high regard at [the company].
qual_6	[The company] is a reliable partner for customers.
qual_7	[The company] is a trustworthy company.
qual_8	I have a lot of respect for [the company].
Performance (PERF)	
perf_1	[The company] is a very well-managed company.
perf_2	[The company] is an economically stable company.
perf_3	The business risk for [the company] is modest compared to its competitors.
perf_4	[The company] has growth potential.
perf_5	[The company] has a clear vision about the future of the company.

Corporate Social Responsibility *(CSOR)*	
csor_1	[The company] behaves in a socially conscious way.
csor_2	[The company] is forthright in giving information to the public.
csor_3	[The company] has a fair attitude toward competitors.
csor_4	[The company] is concerned about the preservation of the environment.
csor_5	[The company] is not only concerned about profits.
Attractiveness *(ATTR)*	
attr_1	[The company] is successful in attracting high-quality employees.
attr_2	I could see myself working at [the company].
attr_3	I like the physical appearance of [the company] (company, buildings, shops, etc.).

The measurement approach has been validated in different countries and applied in various research studies (e.g., Eberl & Schwaiger, 2005; Raithel & Schwaiger, 2014; Raithel, Wilczynski, Schloderer, & Schwaiger, 2010; Schloderer, Sarstedt, & Ringle, 2014). Research has shown that, compared to alternative reputation measures, the approach performs favorably in terms of convergent validity and predictive validity (Sarstedt et al., 2013). The data set used for all analyses in this book stems from Hair et al. (2017) and has 344 observations.

PLS-SEM Software

To establish and estimate PLS path models, users can choose from a range of software programs. A popular early example of a PLS-SEM software program is PLS-Graph (Chin, 1994), which is a graphical interface to Lohmöller's (1987) LVPLS, the first PLS software. Compared to LVPLS, which required the user to enter commands via a text editor, PLS-Graph represents a significant improvement, especially in terms of user-friendliness. However, PLS-Graph has not been further developed in recent years. The same holds for several other early software programs such as VisualPLS (Fu, 2006) and SPAD-PLS (Test&Go, 2006); see Temme, Kreis, and Hildebrandt (2010) for a review. With the increasing dissemination of PLS-SEM in a variety of disciplines, several other programs with user-friendly graphical interfaces were introduced to the market, such as XLSTAT's PLSPM package, Adanco (Henseler & Dijkstra, 2016); PLS-GUI (Hubona, 2015); and particularly SmartPLS (Ringle et al., 2015; Ringle et al., 2005). Finally, users with experience in the statistical software environment R can also draw on packages such as semPLS (Monecke & Leisch, 2012) and plspm (Sánchez, Trinchera, & Russolillo, 2015), which facilitate flexible analysis of PLS path models. To date, SmartPLS 3 is the most comprehensive and advanced program in the field and serves as the basis for all case study examples in this book. The student version of the software is available free of charge at www.smartpls.com. The student version offers practically all functionalities of the full version but is restricted to data sets with a maximum of 100 observations. However, as the data set used in this book has more than 100 observations (344 to be precise), you should use the professional version of SmartPLS, which is available as a 30-day trial version at www.smartpls.com. After the trial period, a license fee applies. Licenses are available for different periods of time (e.g., 2 months, 1 year, 3 years) and can be purchased through the SmartPLS website. The SmartPLS website includes the download area for the software, including the old SmartPLS 2 (Ringle et al., 2005) version, and many additional resources such as short explanations on PLS-SEM and software-related topics, a list of recommended literature, answers to frequently asked questions, tutorial videos for getting started using the software, and the SmartPLS forum, which allows you to discuss PLS-SEM topics with other users.

Setting Up the Model in SmartPLS

Before you draw your model in SmartPLS 3, you need to have data that serve as the basis for running the model. The data we use with the reputation model can be downloaded either as comma-separated value (.csv) or text (.txt) data set in the download section at www.pls-sem.com. Click on **Save Target As . . .** to save the data to a folder on your hard drive and then **Close**. SmartPLS can use both data file formats (i.e., .csv or .txt). Follow the onscreen instructions to save one

Exhibit 1.8   **Corporate Reputation Model**

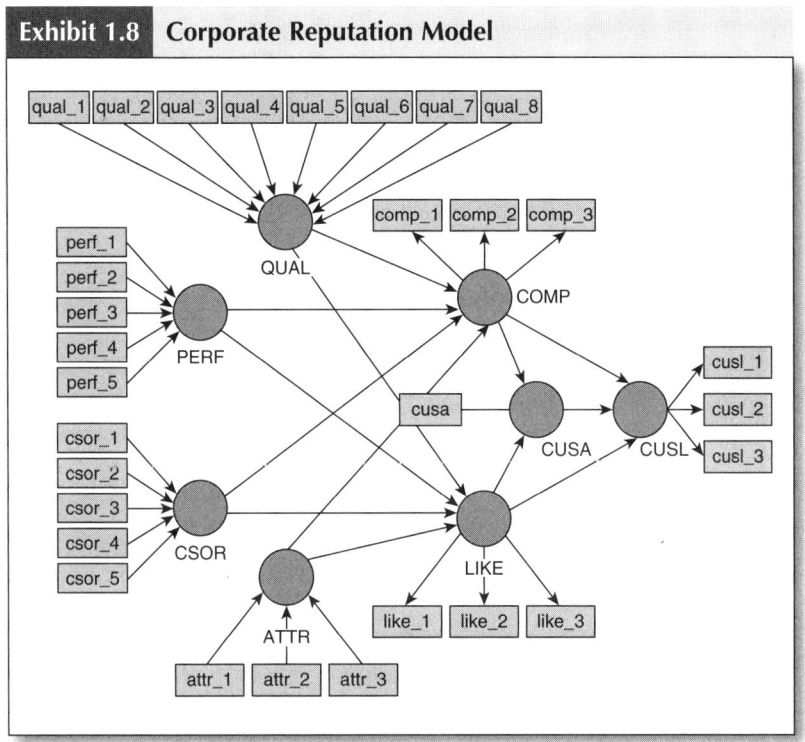

of these two files on your hard drive. With this data, as explained in Chapter 2 of the book by Hair et al. (2017), you can create the PLS path model as shown in Exhibit 1.8.

Alternatively, you can download the ready-to-use corporate reputation project file (**Corporate Reputation – Advanced PLS-SEM Book. zip**) for SmartPLS from www.pls-sem.com. Save the project file on your computer (or other device). Now run the SmartPLS 3 software by clicking on the desktop icon that is available after the software installation on your computer. Another possibility is to go to the folder where you installed the SmartPLS software on your computer. Click on the file that runs SmartPLS and then on the **Run** tab to start the software. To import the corporate reputation project file into the SmartPLS software, use the **File** → **Import Project from Backup File** option in the SmartPLS menu bar. Then you will see the corporate reputation project in the SmartPLS **Project Explorer**. Unfold the project as shown in Exhibit 1.9 and double-click on the **Extended model**. Then the PLS path model as shown in Exhibit 1.8 appears in the SmartPLS modeling window.

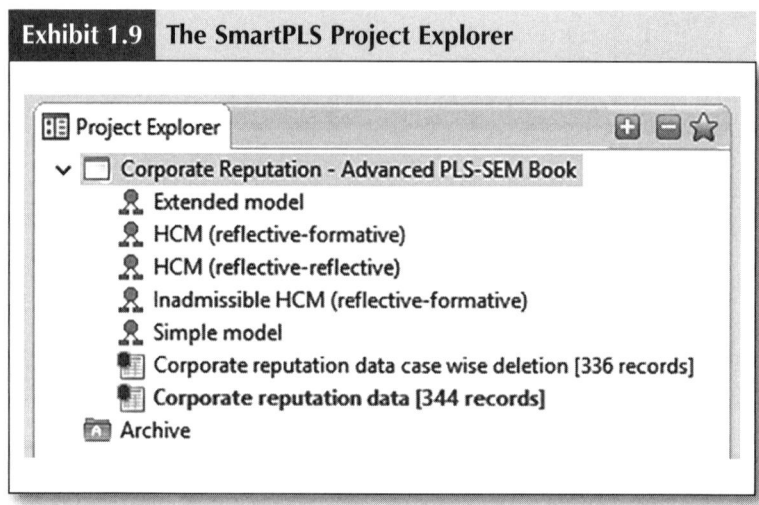

| Exhibit 1.9 | The SmartPLS Project Explorer |

Following the systematic procedure for applying PLS-SEM presented in Hair et al. (2017), the next steps entail the evaluation of the reflective and formative measurement models, followed by an assessment of the structural model. Readers are advised to consult the *Primer on Partial Least Squares Structural Equation Modeling (PLS-SEM)* (Hair et al. 2017) for a detailed discussion and illustration of these analysis steps. The case study illustrations in the following chapters will depart from here, assuming that the quality of the original model's measurement and structural models has been established.

SUMMARY

- **Understand the origins and evolution of PLS-SEM.** The precursors to the PLS-SEM method were two iterative procedures (i.e., principle component regression and PLS-R) that used least squares estimation to develop solutions for single and multicomponent models and for canonical correlation. Further development of these procedures by Herman Wold led to the NIPALS algorithm and a revised generalized version of the PLS algorithm that focused on finding latent variables. In the 1980s Herman Wold proposed his "soft model basic design" underlying PLS-SEM as an alternative to CB-SEM, which has been labeled as hard modeling because of its much more rigorous assumptions in terms of data distribution and sample size (e.g., Falk & Miller, 1992). While both approaches were developed about the same time, CB-SEM became much more widely applied because of its early availability through the LISREL software since the late 1970s. It was not until the debut of Wynne Chin's PLS-Graph software in the mid-1990s that PLS-SEM began developing. With the release of SmartPLS 2 in 2005, PLS-SEM use grew exponentially.

- **Comprehend the principles of and recent developments in measurement.** Conceptual variables represent broad ideas or thoughts about abstract concepts that researchers establish and propose to measure in their research by means of constructs. Based on a construct definition, measurement models express how to measure the construct by means of a set of indicators. Measurement models can be specified reflectively, using effect (i.e., reflective) indicators, or formatively, using causal or composite indicators. Whereas constructs measured with causal indicators have an error term, this is not the case with composite indicators, which define the construct in full. Traditionally, composite indicators have been viewed as a means to combine several variables to represent some new entity whose meaning is defined by the choice of indicators. However, more recent research contends that composite indicators can be used to measure any type of property to which the focal concept refers, including attitudes, perceptions, and behavioral intentions. All measures of conceptual variables are approximations of or proxies for conceptual variables, independent from how they were derived. Thus, irrespective of the quality with which a conceptual variable is theoretically substantiated and operationally defined, and the rigor that encompasses measurement model development, because latent variables stem from data that are inherently imperfect, any measurement in structural equation models produces only proxies for latent variables.

- **Get to know the essential differences between PLS-SEM and CB-SEM and understand when to use each method.** PLS-SEM emphasizes prediction while simultaneously relaxing the demands regarding the data and specification of relationships. PLS-SEM maximizes the endogenous latent variables' explained variance by estimating partial model relationships in an iterative sequence of ordinary least squares regressions. In contrast, CB-SEM estimates model parameters so that the discrepancy between the estimated and sample covariance matrices is minimized. A crucial conceptual difference between PLS-SEM and CB-SEM relates to the way each method treats the latent variables included in the model. CB-SEM considers the constructs as common factors, whereas PLS-SEM follows a composite model perspective using weighted composites of indicator variables to represent the constructs. Importantly, the proxies produced by PLS-SEM are not assumed to be identical to the constructs that they measure—just as it is the case with construct measures produced by CB-SEM. PLS-SEM is not constrained by identification issues, even if the model becomes complex—a situation that typically restricts CB-SEM use. Moreover, PLS-SEM can better handle formatively specified measurement models. Researchers should consider the two SEM approaches as complementary and apply the SEM technique that best suits their research objective, data characteristics, and model setup.

REVIEW QUESTIONS

1. Who developed the generalized PLS-SEM algorithm and what was the intention behind its development?

2. What is the difference between PLS-R and PLS-SEM?

3. What is the difference between common factor models and composite models?

4. What is the difference between reflective and formative measurement models?

5. What is PLSc and what was the intention behind its development?

CRITICAL THINKING QUESTIONS

1. Why is PLS-SEM the preferred method over CB-SEM for prediction?

2. Please comment on the following statement: "Indicators in formative measurement models are error-free."

3. What is the difference between causal and composite indicators?

4. What are the benefits of using PLSc over CB-SEM and PLS-SEM?

KEY TERMS

Attribute

Average variance extracted (AVE)

Causal indicators

CB-SEM

Common factor model

Common variance

Composite indicators

Composite model approach

Conceptual variables

Consistent PLS (PLSc)

Construct definition

Constructs

Covariance-based structural equation modeling (CB-SEM)

Effect indicators

Error variance

Factor (score) indeterminacy

Factor-based SEM

Focal object

Formative measurement model

Indicators

Items

Latent variable

Latent variable scores

Manifest variables

Measurement model

Mimic model

Multiple indicators and multiple causes (MIMIC) model

Partial least squares algorithm

Partial least squares path modeling

Partial least squares regression (PLS-R)

Partial least squares structural equation modeling

PLS algorithm

PLSc

PLS-R

PLS-SEM

Principal components regression

Reflective indicators

Reflective measurement model

Specific variance

Theoretical model

Unique variance

SUGGESTED READINGS

Bollen, K. A. (2011). Evaluating effect, composite, and causal indicators in structural equation models. *MIS Quarterly, 35*, 359–372.

Bollen, K. A., & Diamantopoulos, A. (2017). In defense of causal-formative indicators: A minority report. *Psychological Methods*. Advance online publication. doi:10.1037/met0000056

Hair, J. F., Hult, G. T. M., Ringle, C. M., & Sarstedt, M. (2017). *A primer on partial least squares structural equation modeling (PLS-SEM)* (2nd ed.). Thousand Oaks, CA: Sage.

Hair, J. F., Hult, G. T. M., Ringle, C. M., Sarstedt, M., & Thiele, K. O. (2017). Mirror, mirror on the wall. A comparative evaluation of component-based structural equation modeling methods. *Journal of the Academy of Marketing Science*. Advance online publication. doi: 10.1007/s11747-017-0517-x

Rigdon, E. E. (2012). Rethinking partial least squares path modeling: In praise of simple methods. *Long Range Planning, 45*, 341–358.

Rigdon, E. E. (2016). Choosing PLS path modeling as analytical method in European management research: A realist perspective. *Long Range Planning, 34*, 598–605.

Sarstedt, M., Hair, J. F., Ringle, C. M., Thiele, K. O., & Gudergan, S. P. (2016). Measurement issues with PLS and CB-SEM: Where the bias lies! *Journal of Business Research, 69*, 3998–4010.

CHAPTER 2

Advanced Modeling

CHAPTER PREVIEW

With the rising complexity of theories and cause-effect models in the social sciences, researchers have increasingly used hierarchical component models in their partial least squares structural equation modeling (PLS-SEM) studies (e.g., Ringle, Sarstedt, & Straub, 2012). Hierarchical component models refer to a construct measured at more than one level of abstraction in a PLS path model. More precisely, a hierarchical component model represents a more general construct, measured at a higher level of abstraction, while

37

simultaneously including several subcomponents, which cover more concrete traits of the conceptual variable represented by this construct. Hierarchical component models permit reducing the number of structural model relationships, making the PLS path model more parsimonious, while increasing the bandwidth of content covered by certain constructs (e.g., Jonson, Rosen, & Chang, 2011). In this chapter, we describe the nature of hierarchical component models and discuss how to develop and validate them in a PLS-SEM context.

Additionally, this chapter addresses another advanced modeling matter that concerns nonlinear relationships. Standard conceptualizations that underpin cause-effect relationships in PLS path models imply that constructs affect another in a linear manner. In some instances, however, this assumption does not hold in that relationships may be nonlinear. When the relationship between two constructs is nonlinear, the size of the effect between two constructs depends not only on the magnitude of change in the exogenous construct but also on its value. In analyzing nonlinear effects (Wold, 1982), researchers have to make an assumption regarding the nature of the effect. While an abundance of different effect types is possible, quadratic effects are most common. This chapter deals with this effect type in PLS-SEM and describes its implementation and assessment using SmartPLS 3.

HIERARCHICAL COMPONENT MODELS

Terminology and Motivation

The primer on PLS-SEM (Hair, Hult, Ringle, & Sarstedt, 2017) deals with first-order constructs, which consider a single layer of abstraction. In some instances, however, the constructs that researchers wish to examine are quite complex and can also be operationalized at higher levels of abstraction. Establishing **higher-order models** or **hierarchical component models** (**HCMs**), as they are usually referred to in the context of PLS-SEM (Lohmöller, 1989), most often involve testing **second-order constructs** that contain two layered structures of constructs. For example, satisfaction may be measured at two levels of abstraction. An ensuing HCM would include a general satisfaction construct along with several subconstructs that capture different more concrete attributes of satisfaction, such as satisfaction with the price, satisfaction with the service quality, satisfaction with the personnel, and satisfaction with the servicescape. These more concrete lower-order attributes may form the more abstract satisfaction construct, as shown in Exhibit 2.1.

Instead of modeling the attributes of satisfaction as drivers of the respondent's overall satisfaction and other target constructs (e.g., customer loyalty) in a more complex PLS path model based on a single construct layer, higher-order modeling involves simultaneously mapping the **lower-order constructs (LOCs)** and a single **higher-order construct (HOC)**. Theoretically, this process can be extended to any number of layers yielding third-, fourth-, or fifth-order models, but researchers usually restrict their modeling to two layers (i.e., second-order models). Exhibit 2.2 shows a second example of an HCM of corporate reputation. Different from the previous example, the LOCs likeability and competence represent more concrete manifestations of the HOC corporate reputation. That is, the relationship is from the HOC to the LOCs.

There are several reasons for the inclusion of HCMs in PLS-SEM (e.g., Edwards, 2001; Johnson et al., 2011; Polites, Roberts, &

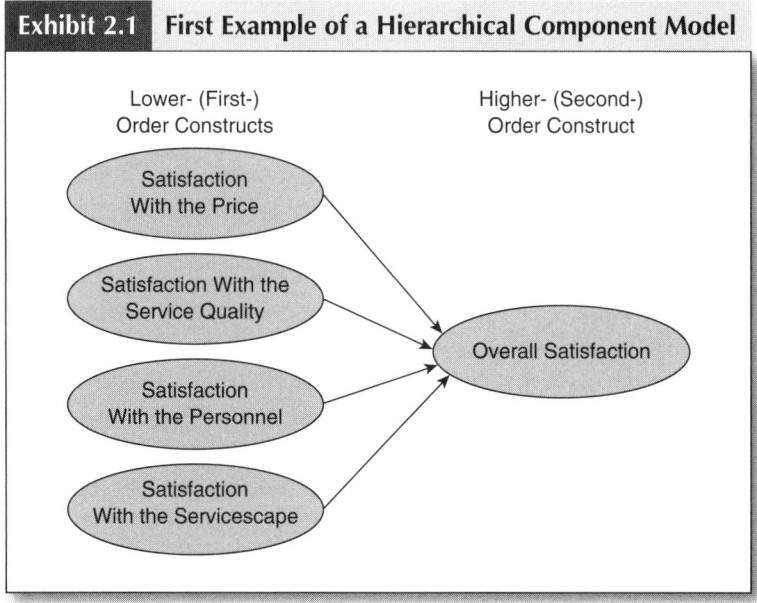

Exhibit 2.1   **First Example of a Hierarchical Component Model**

Lower- (First-) Order Constructs

Higher- (Second-) Order Construct

Satisfaction With the Price

Satisfaction With the Service Quality

Satisfaction With the Personnel

Satisfaction With the Servicescape

Overall Satisfaction

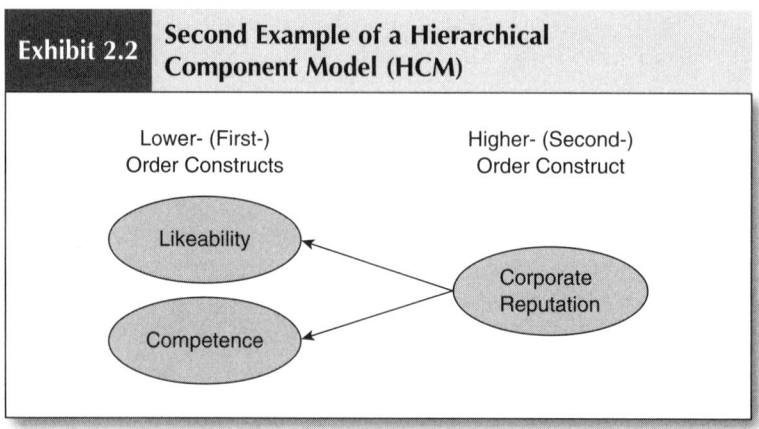

Exhibit 2.2 **Second Example of a Hierarchical Component Model (HCM)**

Thatcher, 2012). One reason pertains to the **bandwidth-fidelity tradeoff,** or the idea that broader constructs are better predictors of criteria that span multiple domains and periods of time. That is, if the goal is to predict broadly defined behaviors, then higher-order multidimensional constructs may prove valuable. Another reason is to overcome the **jangle fallacy,** which occurs when a single phenomenon is examined separately under the guise of two or more variables with different labels.

From a practical perspective, HCMs allow researchers to reduce the number of relationships in the structural model, making the PLS path model more parsimonious and easier to apprehend. Exhibit 2.3 illustrates this aspect. As can be seen in the complex model, there are nine structural model relationships linking the exogenous latent variables (Y_1, Y_2, and Y_3) with the endogenous latent variables (Y_4, Y_5, and Y_6). By including an HOC, the number of path coefficients can be reduced to six, yielding a more parsimonious model in terms of structural model relationships. In this case, the HOC is assumed to fully mediate the LOCs' effects on the endogenous latent variables (for more detail on mediation, see Hair et al., 2017). This reduction in model complexity comes at the expense of predictive power with respect to the endogenous latent variables that the HOC predicts (i.e., Y_4, Y_5, and Y_6 in Exhibit 2.3). The reason is that different from a direct effects model where all exogenous latent variables predict one endogenous latent variable (Exhibit 2.3, left panel), in an HCM, only the HOC predicts the endogenous latent variables (Exhibit 2.3, right panel).

Exhibit 2.3	HCMs and Model Complexity

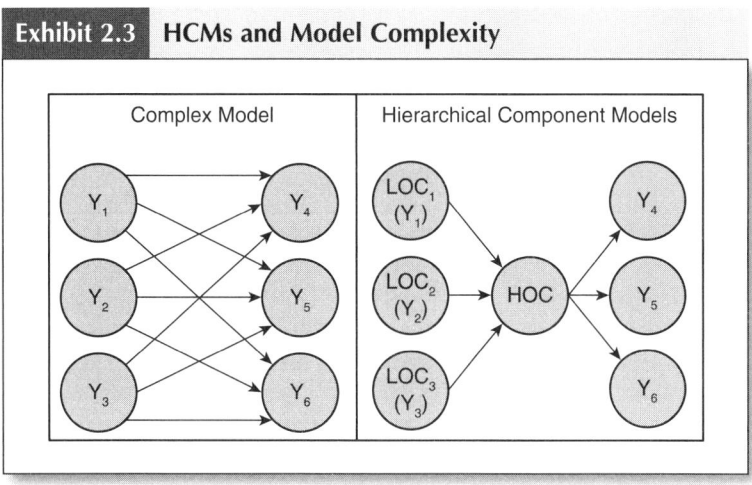

Finally, HCMs also prove valuable if formative indicators in a latent variable's measurement model exhibit high levels of collinearity. This is a crucial issue because it has an impact on the estimates of the weights and their statistical significance. More precisely, high collinearity can result in biased weights and their signs being reversed. Furthermore, collinearity increases standard errors and thus reduces the ability to demonstrate that the estimated weights are significantly different from zero (Hair et al., 2017). HCMs allow handling collinearity problems by offering a vehicle to rearrange measurement models. Provided that measurement theory supports this step, researchers can split up the set of indicators and establish separate constructs in a higher-order structure. Consider, for example, a formatively measured construct with four indicators (x_1-x_4), of which x_1 and x_2 as well as x_3 and x_4 are highly correlated. If conceptually meaningful, researchers could split up the formative construct in two subconstructs, each one being measured with noncollinear indicators (e.g., x_1 and x_3 on the one hand and x_2 and x_4 on the other).

Conceptual Considerations

Establishing HCMs requires researchers to develop and use an appropriate operational definition of the conceptual variable under consideration. The operational definition allows conceptualizing an abstract idea so that it represents the realm of measurable, observable quantities that can be studied (see Chapter 1). The operational definition guides the identification of relevant LOCs of which each refers to a distinctive element (or component) that is associated with the HOC and each has a set of indicators that can be specified by the distinctiveness of the element that characterizes the LOC. At the same time, this characterizing distinctiveness should also be sufficiently relevant so that only those LOCs that are important for the specific study are captured in an HCM. The operational definition, with its characterizing elements, can vary from study to study since a theoretical concept is not per se determined as multidimensional or unidimensional. Rather, a concept can be specified either way, representing different levels of theoretical abstraction (Bollen, 2011).

Conceptually, HCMs can be established following a bottom-up (i.e., inductive) or top-down (i.e., deductive) approach (e.g., Johnson et al., 2011). In the **bottom-up approach**, several latent variables are combined into a single, more abstract construct. On the contrary, in the **top-down approach**, a more abstract construct is defined to consist of several components, as is the case in the customer satisfaction example described above (Exhibit 2.1). Even though frequently used in empirical research in an effort to reduce model complexity, we do not recommend simply summarizing information in a more abstract construct. Establishing an HCM in PLS-SEM always involves a loss of information as the indirect effects between the HOCs and the criterion construct(s) via the newly established HOC replace the direct effects between the HOCs and the criterion construct(s). PLS-SEM, however, allows estimating all relationships in a wider nomological net of constructs (i.e., without the HOC) without loss of information. Therefore, when structural theory supports the inclusion of a larger number of constructs, which on the other hand, however, could be summarized in an HOC, justifying their joint consideration in form of an HCM on the grounds of model parsimony only is not sufficient. Instead, HCMs derived in a top-down manner offer researchers additional insights regarding the effects of different components embedded in a specific construct. Here, the researcher's intention is concretizing the effect of such components on other constructs in the model via the HOC.

In addition to using theory to identify inclusion criteria for selecting suitable LOCs, the nature of relations among the LOCs and the HOC must be clarified. An HOC is a general concept that is either represented (in the reflective mode) or constituted (in the formative mode) by its specific components (i.e., the LOCs). If the HOC is reflective, the general concept is manifested in several more specific LOCs. This type of model is also referred to as a **spread model** (Lohmöller, 1989). If the HOC is formative, it is a combination of several specific LOCs representing more concrete components that form the general concept (Becker, Klein, & Wetzels, 2012; Edwards, 2001; Wetzels, Odekerken-Schroder, & van Oppen, 2009). This model type is also referred to as a **collect model**. The HCM in Exhibit 2.1 has formative relationships going from the LOCs to the HOC, representing each LOC's relative contribution to forming the HOC. However, if operationalized differently, these relationships could also have been modeled in the opposite direction with the LOCs reflecting the HOC. A formative specification is appropriate when the operational definition of the conceptual variable suggests that a change in an LOC's value due, for example, to a change in a respondent's assessment of the trait being captured by the LOC changes the value of the HOC. Analogous to indicators in formative measurement models, the LOCs do not need to be, but can be, correlated as they do not represent concrete manifestations of the HOC.

Instead, a reflective specification is appropriate when there is a more general, abstract construct that explains the correlations between the LOCs as shown in Exhibit 2.2. Hence, there should be substantial correlations between the LOCs that—analogous to reflective measurement models—are assumed to be caused by the HOC. That is, the HOC is the spurious cause explaining the correlations between the LOCs.

In addition to the relationship between the HOC and the LOCs, HCMs can be characterized on the grounds of the specification of the LOCs' measurement models (i.e., reflective or formative). As a result, four main types of HCMs evolve (Exhibit 2.4) as discussed in the extant literature (Jarvis, MacKenzie, & Podsakoff, 2003; Wetzels et al., 2009). The **reflective-reflective HCM** (Type I; Exhibit 2.4) indicates a reflective relationship between the HOC and the LOCs, the latter of which are all measured reflectively. In this type of HCM, the LOCs are highly correlated and the HOC represents the spurious cause explaining these correlations. Lohmöller (1989) calls this type of HCM **hierarchical common**

factor model, where the general HOC represents the common factor of several specific factors (i.e., LOCs). Therefore, this type of HCM model is most appropriate if the objective of the study is to establish the common factor of several related yet distinct reflective LOCs (Becker et al., 2012). The use of reflective-reflective HCMs has been subject to considerable debate, with critics arguing that such models do not exist (or are meaningless) since reflective measures should be unidimensional and conceptually interchangeable, which conflicts with the view of multiple underlying dimensions being distinct in nature (Lee & Cadogan, 2013).

Exhibit 2.4	Types of Hierarchical Component Models

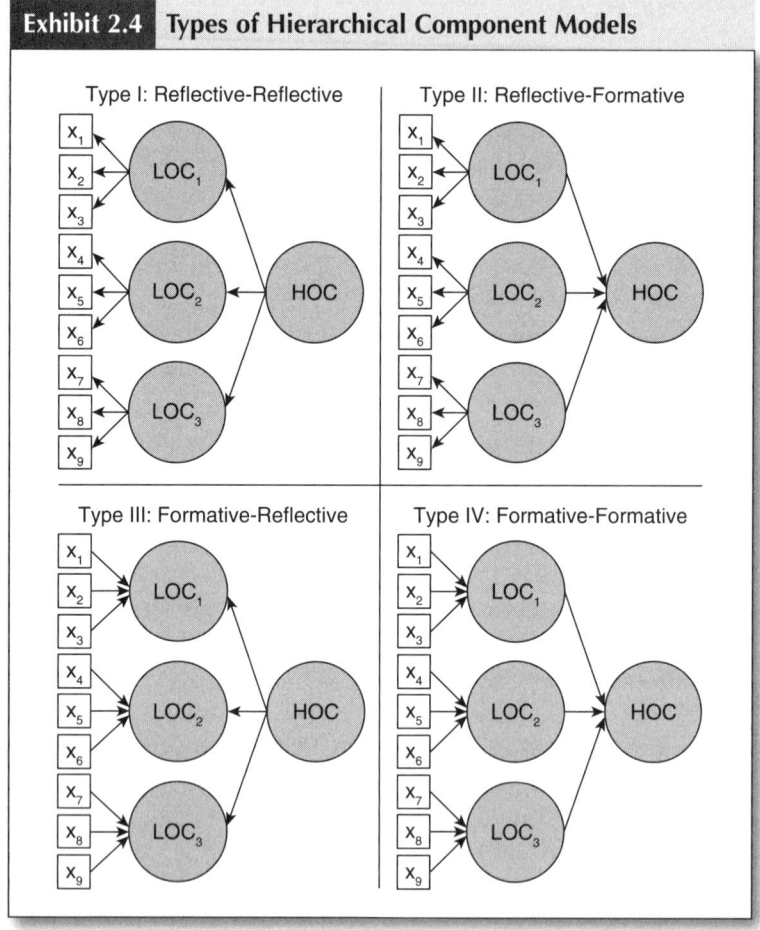

Source: Adapted from Figure B1 in Ringle, Sarstedt, & Straub (2012).

Proponents argue this critical view is conceptually flawed, concluding that researchers should consider this model type as a legitimate operationalization option for multidimensional constructs (Temme & Diamantopoulos, 2016). In particular, researchers can use reflective-reflective HCMs to measure a concept multiple times at different points in time. In such a setting, the LOCs represent the different measurements of a concept at different points in time (i.e., different batteries of a test sequence), which the HOC explains simultaneously. If the test batteries (i.e., the LOCs) measure the same construct at different times, they are highly correlated. Lohmöller (1989) characterizes this constellation as a **multiple battery model** and presents the example of the general ability of schoolchildren. Each LOC represents a test battery of verbal, numerical, and spatial indicators measured at different points of time (e.g., tests at the beginning, middle, and end of the school year) for the same class (i.e., the same individuals). However, the multiple battery model appears to not have been applied in PLS-SEM studies thus far.

In a **reflective-formative HCM** (Type II; Exhibit 2.4), the HOC represents a more general construct of the reflectively measured LOCs. The specific LOCs do not necessarily share a common cause but rather form the general HOC. Barroso and Picón (2012) offer an example of a reflective-formative HCM in their multidimensional analysis of perceived switching costs (see Chapter 1). They identify a set of six dimensions (benefit loss costs, personal relationship loss costs, economic risks costs, evaluation costs, setup costs, and monetary loss costs) that represent LOCs of the more general HOC perceived switching costs. "A modification in one dimension does not necessarily imply a modification in another. In other words, they do not necessarily covary; rather, each dimension can vary independently of the others" (Barroso & Picón 2012, p. 532). For this reason, unlike some prior research, Barroso and Picón (2012) propose that perceived switching costs is an aggregate construct that is expressed as a composition of its different LOCs (see Becker et al. [2012] for further examples of reflective-formative HCMs). Within a larger nomological net, the HOC fully mediates the relationships of the LOCs with the endogenous latent variables in the PLS path model. For example, in the HCM in Exhibit 2.3, the HOC fully mediates the relationship between LOC_1 and the three endogenous constructs Y_4, Y_5, and Y_6.

Another HCM is the **formative-reflective** type (Type III; Exhibit 2.4). The formative-reflective HCM includes a more general HOC that explains the formatively measured LOCs. The objective of this HCM type is extracting the common part of several formatively measured LOCs that have been established to represent the same theoretical content. However, every LOC builds on a set of different indicators. By using several formatively measured LOCs, researchers can overcome the problem that a stand-alone construct measured with formative indicators can hardly cover the construct domain in full. Using similar yet distinct formatively measured LOCs as representations of the HOC offers a broader coverage of the construct domain (Becker et al., 2012). A typical example of this HCM type is overall firm performance in which several formatively measured LOCs embody performance-relevant characteristics. The HOC represents the common part of the LOCs (i.e., overall firm performance in the HCM; Jarvis et al., 2003; Petter, Straub, & Rai, 2007). Alternatively, the formative-reflective HCM can serve as a multiple battery model as explained in the context of the reflective-reflective HCM. In that case, the LOCs represent the same construct that has been formatively measured with the same indicators and for the same observations at different points of time. In this type of multiple battery model, the relationships from the HOC to the LOCs will be of similar magnitude as these represent one concept measured at different points in time.

Finally, the **formative-formative HCM** (Type IV; Exhibit 2.4) determines the relative contribution of the formatively measured LOCs to the more abstract HOC. This HCM type proves useful to structure a complex formative construct with many indicators into several subconstructs, as is the case when researchers subsume several concrete aspects under a more general concept. Again, firm performance would represent a concept of this nature to be measured using this HCM type. While the formative-reflective HCM type would comprise different indices of overall firm performance as LOCs, the formative-formative HCM type includes LOCs representing different aspects of performance such as the performance of different organizational activities or subdivisions (e.g., R&D performance, HR performance, sales performance) that determine overall firm performance (i.e., the HOC; Jarvis et al., 2003; Petter et al., 2007) but that do not necessarily have to correlate. Another example of a formative-formative HCM is the marketing mix as determined by the four Ps (i.e., price, product, promotion, and place).

Although some hierarchical latent variable model types are less likely to be used, all of these types can be found in empirical research. Ringle et al. (2012) report that out of 109 PLS-SEM studies in the management information systems flagship journal *MIS Quarterly*, 15 studies (23%) reported 25 HCMs. Of these 25 models, the reflective-formative HCM Type II (13 models; 52%) and formative-formative HCM Type IV (six models; 24%) were most frequently employed in these empirical applications. Hence, researchers seem to predominantly apply HCM when modeling the formative relationships between the LOCs and the HOC. Moreover, in most studies (13 studies; 87%) the HOC was embedded in a nomological network of latent variables serving as antecedent (five studies; 33%), consequence (three studies; 20%), or both (five studies; 33%). Therefore, the discussion should not be limited to HCMs as separate constructs, also referred to as **stand-alone HCMs**, but should also consider their potential application in a nomological network of latent variables embedded in a structural model (Becker et al., 2012). Furthermore, the examples of HCMs discussed in this section are illustrative only such that the same concepts, when defined differently within divergent theoretical nomological networks, may need to be modeled drawing on a different HCM type than described here.

Technical Considerations

Overview

PLS-SEM allows the specification and estimation of all HCM types, as shown in Exhibit 2.4. The specific type of HOC demands careful consideration when specifying and estimating the model, however, since PLS-SEM requires each construct in the PLS path model to have at least one indicator in its measurement model. This necessity holds not only for LOCs but also for the HOC, which is an abstract representation of the LOCs. As such, the nature of an HOC is different from that in covariance-based SEM (CB-SEM), where an HOC uses other latent variables for measurement and has no indicators in its measurement model. For this reason, HOCs are sometimes called phantom variables in CB-SEM.

To handle the measurement issue of HCMs in PLS-SEM, researchers can draw on two main approaches for their specification in a PLS path model (Becker et al. 2012; Ringle et al., 2012; Wetzels et al., 2009): (1) the repeated indicators approach and (2) the two-stage approach. After introducing these two approaches, we discuss some additional technical considerations and offer recommendations.

The Repeated Indicators Approach

In the repeated indicators approach, all indicators of the LOCs are assigned to the measurement model of the HOCs (Lohmöller, 1989; Wold, 1982). In the second-order model examples in Exhibit 2.4, the repeated indicators approach uses the indicators x_1 to x_9 of the LOCs to establish the measurement model of the HOC as shown in Exhibit 2.5. Consequently, the indicators are used twice: once for the LOCs (primary loadings/weights) and again for the HOC (secondary loadings/weights). This approach can easily be extended to third- or fourth-order HCMs (e.g., Wetzels et al., 2009),

Exhibit 2.5   HCMs Using the Repeated Indicators Approach

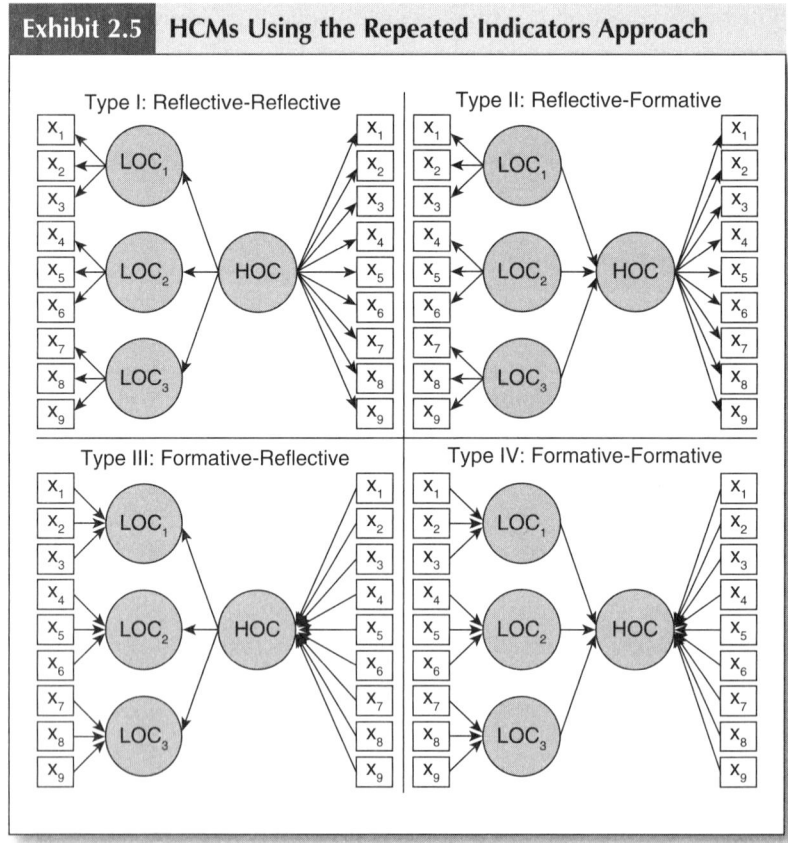

Source: Adapted from Figure B1 in Ringle, Sarstedt, & Straub (2012).

if theoretically and conceptually meaningful. Exhibit 2.5 shows how to establish the four HCM types using the repeated indicators approach.

As described in Chapter 1, PLS-SEM draws on two modes to estimate construct measures, which relate to the way the method estimates the indicators' weights, representing each indicator's contribution to the composites. Mode A corresponds to correlation weights derived from bivariate correlations between each indicator and the construct. Mode B corresponds to regression weights, the standard in ordinary least squares regression analysis. Regression weights take into account not only the correlation between each indicator and the construct but also the correlations between the indicators. Mode A is usually associated with the estimation of reflectively measured constructs, whereas Mode B is used to estimate formatively measured constructs. Correspondingly, reflective-reflective and reflective-formative HCMs specified with repeated indicators commonly draw on Mode A, while formative-reflective and formative-formative HCMs generally use Mode B estimation.

Instead of following this reflex-like application of these estimation modes, researchers can also use Mode A for estimating formatively measured constructs and Mode B for estimating reflectively measured constructs (see Chapter 1). Lohmöller (1989) advocates such a broader view on establishing and estimating HCMs by distinguishing 16 different ways to specify and estimate HCMs with reflectively measured constructs, depending on the measurement model estimation of the LOCs and the HOC (Mode A or Mode B), the relationship between the LOCs and the HOC (collect model or spread model), and the inner model weighting scheme (factor vs. path weighting scheme). The author reports two constellations that show **premature convergence**, which means that the PLS-SEM algorithm converges but the results depend on the starting values and are not optimal. These constellations are (1) Mode A estimation for the LOCs and the HOC, relationships from the LOCs to the HOC, and use of path weighting scheme; and (2) Mode A estimation for the LOCs, Mode B estimation for the HOC, relationships from the HOC to the LOCs, and use of path weighting scheme. Becker et al. (2012) substantiate these results drawing on a complex simulation study and show that the path weighting scheme yields the greatest parameter biases when using the standard model estimation (i.e., Mode A for reflectively measured constructs) in a reflective-formative HCM. In these settings, the factor and centroid weighting schemes—the latter of which Lohmöller did not consider in his review of HCM types—performed much better in terms of parameter accuracy. However, future research is needed to provide insights that substantiate the appropriate choice of inner weighting scheme (centroid, factor, or path) when using HCMs in PLS-SEM. Against this background, we suggest using the factor weighting scheme as a compromise solution between the centroid and path weighting schemes when a PLS path model includes one or more HCMs.

Similarly, clear understanding is missing about when one should deviate from the standard practice of using Mode A estimation for reflectively specified constructs and Mode B for formatively specified constructs in an HCM. While Becker et al. (2012) recommend the use of Mode B for estimating a reflectively specified HOC in a reflective-formative HCM, it remains unclear whether this suggestion generalizes to other HCM types. Therefore, we advise reverting to standard practice until further research specifies situations that favor the use of Mode A and Mode B.

The repeated indicators approach warrants additional technical attention in three respects. First, the relationships between the HOC and LOCs are significantly biased if the LOCs' measurement models vary strongly in terms of number of indicators (Becker et al., 2012). Consider, for example, an HCM with two LOCs, one with eight indicators (LOC_1) and another one with two indicators (LOC_2). When using the repeated indicators approach, all 10 indicators are simultaneously assigned to the HOC. As they share a large number of indicators in the HCM, a stronger relationship between the HOC and LOC_1 is expected. Furthermore, the predictive power of the LOCs is partly a function of the reliability, which depends on the number of indicators (Diamantopoulos, Sarstedt, Fuchs, Kaiser, & Wilczynski, 2012). More indicators entail higher reliability and, hence, stronger predictive power (Peter & Churchill, 1986), particularly when the measurement model is reflective. To avoid any model-induced biases in the parameter estimates, researchers should ensure that the number of indicators is equal (or at least comparable) across the LOCs. If this is not feasible from a measurement theory perspective, researchers should systematically assess the effect of indicator elimination on the relationships between the HOC and the LOCs. This particularly applies when the LOCs are measured formatively as formative measurement models often entail differences in measurement model complexity, depending on the conceptual domain's breadth. Note that if indicators of any LOCs are removed, the researcher must ensure that the same indicators are also removed from the HOC. Otherwise, bias will be introduced into the model estimates.

Second, researchers should evaluate not only the measurement models of the LOCs but also the measurement model of the HOC. However, different from other constructs in the PLS path model, the assessment of the HOC is not concerned with the relationships between the HOC and its indicator variables but the relationships between the HOC and its LOCs. While these relationships are mapped as path coefficients in a PLS-SEM analysis, from a modeling perspective they correspond to loadings (in case of reflective-reflective and formative-reflective HCMs) or weights (in case of reflective-formative or formative-formative HCMs) and need to be interpreted as such. That is, in reflective-reflective and formative-reflective HCMs, the relations between the HOC and the LOCs serve as input for the estimation of the average variance extracted (AVE), composite reliability, and Cronbach's alpha. Discriminant validity between the HOC and the LOCs is not of concern as conceptual and empirical redundancies are expected, rendering discriminant validity assessment between these model elements meaningless. However, discriminant validity needs to be established between the LOCs in a particular HCM. For this kind of assessment, use the heterotrait-monotrait ratio of correlations (HTMT) criterion (Henseler, Ringle, & Sarstedt, 2015). Whether one needs to examine the discriminant validity between the HOC and other constructs in the model is subject to controversy. In case you want to examine the discriminant validity of the HOC, you need to compute the HTMT criterion manually (see Henseler et al., 2015) using the construct correlations among the LOCs as monotrait-heteromethod correlations, and the indicator cross-loadings of constructs outside the HCM with the LOCs as heterotrait-heteromethod correlations. Finally, in reflective-formative or formative-formative HCMs, researchers also need to assess collinearity as well as significance and relevance of the relations between the LOCs and the HOC.

Third, the use of reflective-formative and formative-formative HCMs is problematic when the HOC has an antecedent latent variable(s). As illustrated in Exhibit 2.6, such a constellation entails that almost all of the HOC's variance is explained by its LOCs, yielding an R^2 value of (close to) 1.0 ($R^2 \approx 1.0$). As a result, any further path coefficients (i.e., other than those by the LOCs) for relationships pointing at the HOC (i.e., P_1 in Exhibit 2.6) will be very small (perhaps zero) and insignificant (Ringle et al., 2012).

Exhibit 2.6	Problem of Using an HCM as an Endogenous Latent Variable in a PLS Path Model

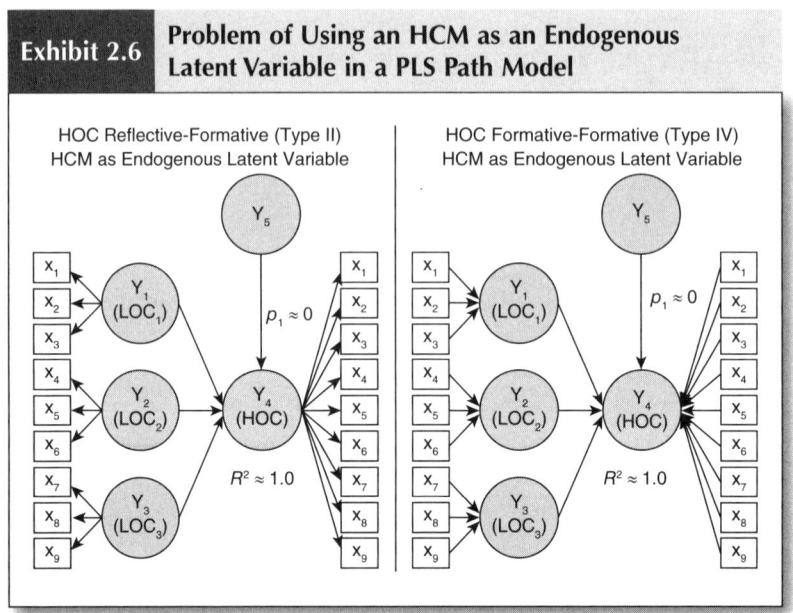

Source: Adapted from Figure B2 in Ringle, Sarstedt, & Straub (2012).

To handle such model constellations, Becker et al. (2012) suggest modeling the impact of the antecedent latent variable(s) not directly on the HCM but via the LOCs as illustrated in Exhibit 2.7. The total effect of the antecedent latent variable (Y_5) on the HOC is equal to its indirect effect via the LOCs (i.e., $p_2 \cdot p_7 + p_3 \cdot p_6 + p_4 \cdot p_5$) plus the direct

Exhibit 2.7	Total Effects Analysis of Collect-Type HCM Using the Repeated Indicators Approach

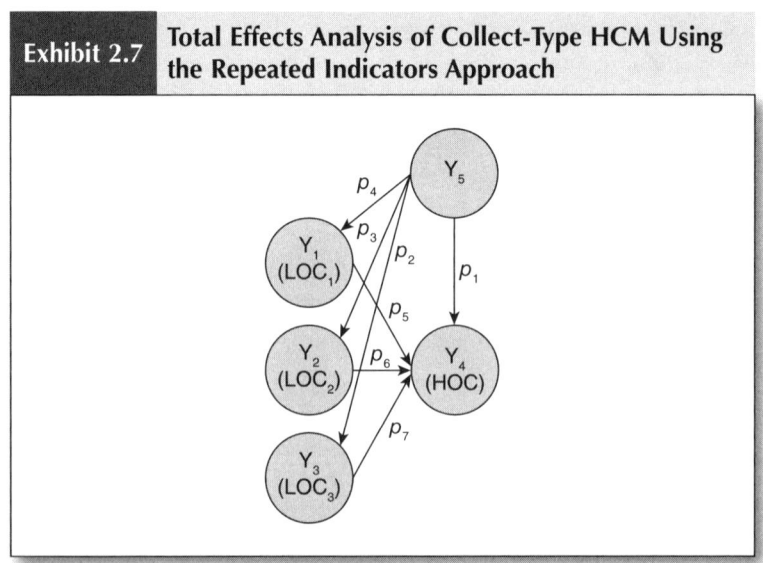

effect of the antecedent latent variable on the HOC (i.e., p_1). That is, researchers need to additionally specify the relationships between the antecedent construct and the LOCs and analyze its total effect on the HOC. This analysis, also referred to as **total effects analysis of collect-type HCMs,** should routinely be applied to avoid Type II errors in the estimation of the direct effect between the antecedent latent variable and the HOC (p_1 in Exhibit 2.6).

The Two-Stage Approach

The **two-stage approach for HCMs** is an alternative to solving the problems that occur when the HOC of reflective-formative type and formative-formative type HCMs is explained by antecedent latent variable(s) that are not its LOCs in the wider nomological net of the PLS path model (Ringle et al., 2012). The approach draws on the same principle as the two-stage approach in moderator analyses in PLS-SEM (Henseler & Chin, 2010) in that it combines the repeated indicators approach with an analysis of latent variable scores. In the first stage, the repeated indicators approach—as shown in Exhibit 2.6—is used to obtain the LOCs' scores. These scores are saved in the data set as additional variables for further analysis in the second stage. Then, in the second stage, the LOCs' scores serve as manifest variables in the HOC's measurement model (Exhibit 2.8).

| Exhibit 2.8 | **Stage Two of the Two-Stage Approach for the HCM Analysis** |

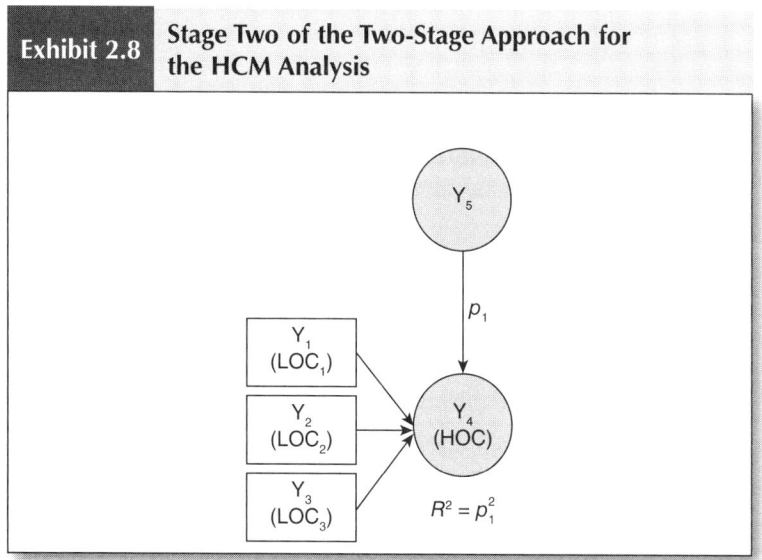

Source: Adapted from Figure B2 in Ringle, Sarstedt, & Straub (2012).

Technically, one estimates the PLS path model of the first stage using the repeated indicators approach (Exhibit 2.6) and adds the latent variable scores as new variables to the data set. Then, one uses this data set and creates the PLS path model of the second stage using the latent variable scores of the first stage as indicators in the HOC's measurement model (Exhibit 2.8). While the approach was originally proposed to handle Type II errors in reflective-formative type and formative-formative type HCMs, the two-stage approach can be generally applied to all HCM types (i.e., also to reflective-reflective and formative-reflective HCMs). Finally, since the latent variable scores of the first stage become the indicators of the HOCs' measurement models, the assessment can draw on the well-established criteria for reflective and formative measurement model assessment (see Hair et al., 2017).

Recommendations and Rules of Thumb

Simulation evidence on the relative advantages of the repeated indicators and two-stage approaches for estimating HCM is scarce and has considered only the reflective-formative type HCM. Becker et al.'s (2012) results suggest that the repeated indicators approach should generally be preferred over the two-stage approach. With the introduction of the total effects analysis of collect-type HCMs, the repeated indicators approach can also be used in settings where constructs other than the LOCs predict the HOC in reflective-formative and formative-formative HCMs. Nevertheless, the two-stage approach proves valuable if the researcher seeks to assess the nature of the higher-order construct using confirmatory tetrad analysis (CTA-PLS; see Chapter 3). Using the LOCs' latent variable scores from stage 1 as input for the measurement model specification in stage 2, the CTA-PLS assesses the covariance structure of the HOC's measurement model to test whether the HCM corresponds to a collect model or spread model. Such an assessment is not feasible with the repeated indicators approach because the CTA-PLS—as implemented in software programs such as SmartPLS 3—considers the covariances only in the measurement models and not among the latent variables. Based on the CTA-PLS results, researchers gain additional insights to specify the HCM and can re-estimate the model using the repeated indicators approach. Exhibit 2.9 summarizes the rules of thumb for HCM analyses in PLS-SEM.

Exhibit 2.9	Rules of Thumb for Hierarchical Component Models

- Clearly examine and describe the theoretical and conceptual foundations of the HCM type.
- Generally, use the repeated indicators approach to model HCMs in a PLS path model. Ensure that the LOCs have an equal number of indicators. If not feasible, assess the impact of indicator deletion on the relationships between the HOC and the LOCs.
- Use the two-stage approach when the aim is to evaluate relationships between the HOC and the LOCs using the CTA-PLS. Use the latent variable scores from stage 1 as input for the measurement model specification in stage 2, and run the CTA-PLS on the model.
- In case the HOC of a reflective-formative or formative-formative HCM has an antecedent construct other than the LOCs, apply the total effects analysis of collect-type HCMs using the repeated indicators approach.
- Generally, use the factor weighting scheme for parameter estimation. When the reflectively measured HOC is estimated using Mode B, use the path weighting scheme.
- Ensure that all constructs in the HCM meet all standard measurement model evaluation criteria. Different from other constructs in the PLS path model, the assessment of the HOC is not concerned with the relationships between the HOC and its indicator variables but the relationships between the HOC and its LOCs. These relationships correspond to loadings (in case of reflective-reflective and formative-reflective HCMs) or weights (in case of reflective-formative or formative-formative HCMs) and need to be interpreted accordingly. Discriminant validity does not need to be established between the HOC and the LOCs. However, the LOCs should show discriminant validity with all other constructs in the PLS path model. Consider examining the discriminant validity between the HOC and other constructs in the model.

Case Study Illustration

Drawing on the case study model and data presented in Chapter 1, we outline how to create an HCM for the construct corporate reputation. The corporate reputation model focuses on the *COMP* and *LIKE* constructs representing two separate dimensions of corporate reputation (Exhibit 2.2). Instead of modeling the distinct impact of the antecedent constructs (i.e., *ATTR*, *CSOR*, *PERF*, and

QUAL) on *COMP* and *LIKE* as well as their effect on the criterion variables (i.e., *CUSA* and *CUSL*) separately, these two constructs could be handled as subdimensions of a more general corporate reputation construct. By establishing a second-order construct with *COMP* and *LIKE* as LOCs, the PLS path model becomes more parsimonious. From a measurement theory perspective, *COMP* and *LIKE* can be considered reflections of corporate reputation (e.g., Schwaiger, 2004), thereby implying the use of a reflective-reflective HCM type since each of the LOCs is measured reflectively. However, others argue that *COMP* and *LIKE* determine corporate reputation (e.g., Eberl, 2010), therefore inferring a reflective-formative HCM type specification. Since we deal with only two LOCs, the CTA-PLS, which would require at least four LOCs (see Chapter 3 for more detail), cannot provide additional empirical substantiation concerning the direction of the relationship between the LOCs and the HOC. The correlation of the LOCs should be high and substantial for a reflective-reflective HCM (e.g., >0.5). Otherwise, the estimated coefficients for the relationships between the HOC and the LOCs, which are interpreted as loadings, will most likely result in values below the desired threshold of 0.7. On the contrary, for reflective-formative HCMs, very high correlations among the LOCs yielding variance inflation factor (VIF) values above 5 are not desirable. When using a reflective-reflective HCM, *COMP* and *LIKE* have a correlation of 0.631, providing support for this model specification. At the same time, when assuming a reflective-formative HCM, the correlation between *COMP* and *LIKE* is 0.633, which yields a VIF value of $1/(1-0.633^2) = 1.669$. Thus, both HCM types can be empirically supported. For this reason, in the following sections, for illustrative purposes, we demonstrate and analyze both approaches separately.

Reflective-Reflective HCM

To establish the reflective-reflective HCM of corporate reputation, we draw on the repeated indicators approach. The constructs *COMP* and *LIKE* represent the LOCs of the more general construct corporate reputation (*REPU*), which is measured with the six indicators *comp_1*, *comp_2*, *comp_3*, *like_1*, *like_2*, and *like_3*. That is, all indicators of the reflectively measured LOCs are simultaneously assigned to the reflective measurement model of the HOC (Exhibit 2.10). In terms of structural model specification, the four antecedent constructs *ATTR, CSOR, QUAL,* and *PERF* directly relate to *REPU* (instead of *COMP* and *LIKE*). Similarly, *REPU* is now directly related to the two criterion variables *CUSA* and *CUSL*.

In order to create the HCM as shown in Exhibit 2.10, navigate to the SmartPLS **Project Explorer,** and right-click on **Extended model** in the corporate reputation project. In the menu that opens, select the **Copy** option. Now right-click again and select the **Paste** option in the menu. A dialogue opens that allows you to enter a name for the copy of the newly added model. Use a self-explaining name such as **HCM (reflective-reflective)**. After pressing the **OK** button, the new model, which is a copy of the corporate reputation model, appears in the **Project Explorer** and the model automatically opens in the modeling window. Instead of using the **Copy** and **Paste** options, you can also use the **Duplicate** option. Next, navigate to the modeling window and select the latent variable *COMP.* Right-clicking on *COMP* opens a menu that includes the **Delete** option. Choose this option and continue by also deleting the *LIKE* construct.

Using the menu options **Edit → Add Latent Variable . . . , Edit → Rename,** and **Edit → Add Connection . . .** or by using the corresponding options in the tool bar, enter three new constructs into the modeling window labeled *COMP, LIKE,* and *REPU.* Next, drag and drop the items from the **Indicators** view on the lower left side of the screen on the latent variables in the modeling window. Make sure that you select the **Show All Indicators** filter (colored in gray) in the **Indicators** view to allow for the repeated assignment of *comp_1, comp_2, comp_3, like_1, like_2,* and *like_3* to the *REPU* construct. The final model should look similar to the PLS path model shown in Exhibit 2.10. An easier and faster way to add a latent variable with its indicators to the modeling window is by pressing the Control key and selecting the indicators of a certain construct (e.g., *comp_1, comp_2, comp_3*). Then, left-click on the selected indicators, hold the mouse button, and move all selected indicators at the same time to the modeling window. After you have dropped the indicators on an empty area in the modeling window, a latent variable appears with the previously selected indicators already assigned. Now move the latent variable to the position where you want it to be, use the **Rename** option, which appears after right-clicking on the latent variable, and use the **Connect** option in the tool bar to create relationships with other latent variables in the PLS path model. Instead of newly creating the HCM model for the corporate reputation example, you can also download the **Corporate Reputation - Advanced PLS-SEM Book** project from www.pls-sem. com. After using the **Import → Import Project From Backup File** option, you can open the **HCM (reflective-reflective)** model in the **Corporate Reputation** project. Then, the PLS path model shown in Exhibit 2.10 appears in the modeling window.

To estimate the PLS path model, click on **Calculate → PLS Algorithm** in the SmartPLS menu. Alternatively, you can left-click on the wheel symbol with the label **Calculate** in the tool bar. Run the PLS-SEM algorithm using the **Factor** weighting scheme and the default settings for the maximum number of iterations and the stop criterion. Make sure to select the **Mean Replacement** option in the **Missing Values** tab, and do not select a weighting vector in the **Weighting** tab. After clicking the **Start Calculation** button, the SmartPLS results report opens automatically, provided the option **After Calculation: Open Full Report** has been selected in the start dialog of the PLS-SEM algorithm.

In the next step, you will need to evaluate the measurement models (Hair et al., 2017). The evaluation results of the formative and reflective measurement models differ only slightly from those presented in Chapters 4 and 5 in Hair et al. (2017). This also holds for the *LIKE* and *COMP* constructs, which now act as LOCs. Specifically, measures of *LIKE* yield satisfactory levels of convergent validity (AVE = 0.747) and internal consistency reliability (composite reliability = 0.899; Cronbach's alpha = 0.831). Similarly, measures of *COMP* exhibit convergent validity (AVE = 0.691) and internal consistency reliability (composite reliability = 0.870; Cronbach's alpha = 0.776). Our primary focus, however, is the newly established HOC, whose measurement model is described by *REPU*'s path relationships with the two LOCs *LIKE* and *COMP*. As you can see in Exhibit 2.10, the HCM's loadings are 0.914 for *LIKE* and 0.891 for *COMP*. As these loadings technically correspond to path coefficients in the PLS path model, SmartPLS does not present measurement model statistics for the HCM. However, you can easily calculate the relevant statistics manually. The AVE is the mean of the HCM's squared loadings; therefore, you can obtain an AVE value of $(0.914^2 + 0.891^2)/2 = 0.815$, indicating convergent validity.

Exhibit 2.10 Reflective-Reflective HCM Example

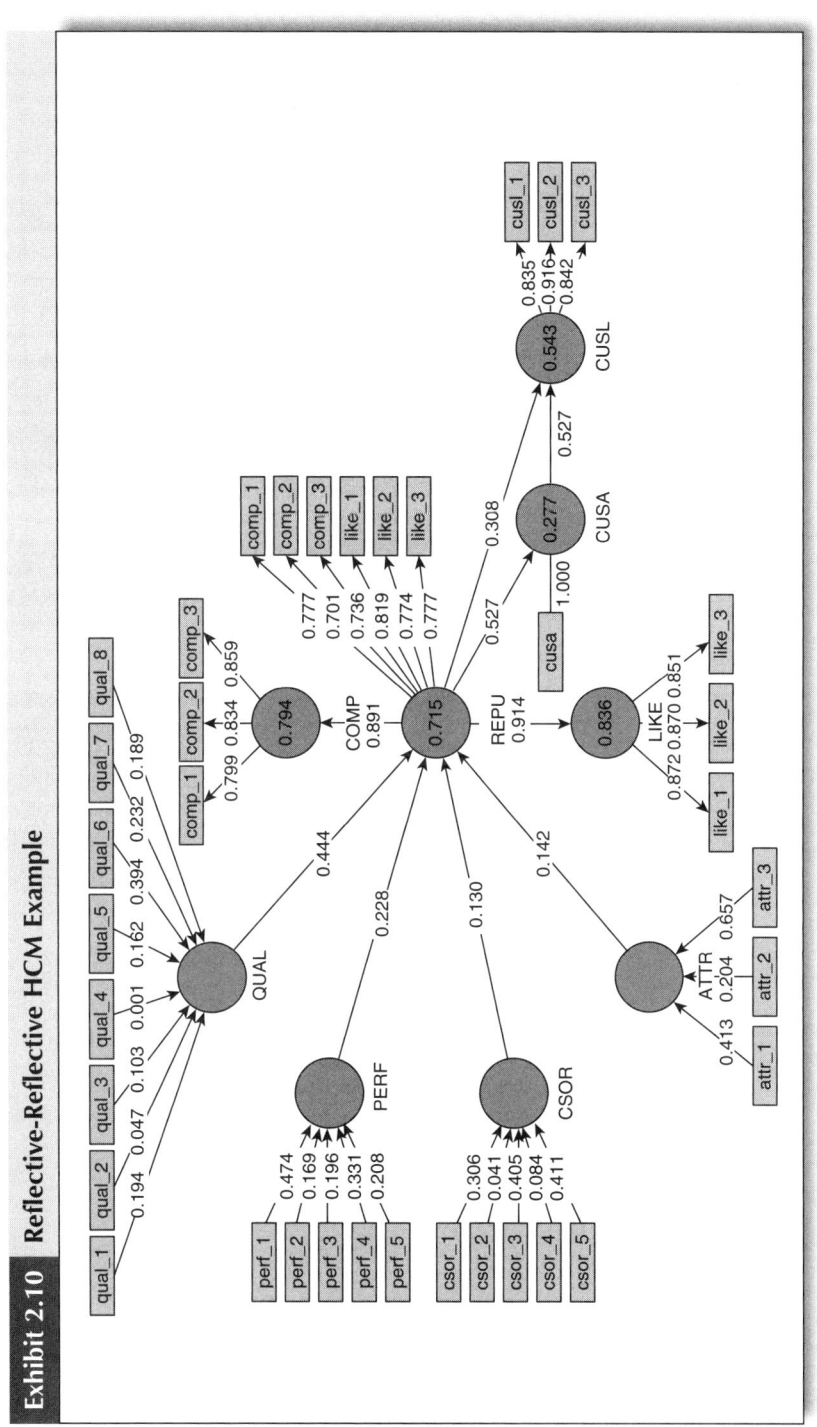

Similarly, you can manually compute the composite reliability using the following formula:

$$\rho_c = \frac{\left(\sum_{i=1}^{M} l_i\right)^2}{\left(\sum_{i=1}^{M} l_i\right)^2 + \sum_{i=1}^{M} \text{var}(e_i)},$$

where l_i symbolizes the loading of the LOC i of a specific HOC measured with M LOCs ($i = 1, \ldots, M$), e_i is the measurement error of LOC i, and var(e_i) denotes the variance of the measurement error, which is defined as $1 - l_i^2$. Entering the two loading values yields the following:

$$\rho_c = \frac{(0.914 + 0.891)^2}{(0.914 + 0.891)^2 + (1 - 0.914^2) + (1 - 0.891^2)}$$

$$= \frac{3.258}{3.258 + 0.206 + 0.164} = 0.898$$

Similarly, in PLS-SEM the (standardized) Cronbach's alpha is given by:

$$\text{Cronbach's } \alpha = \frac{M \cdot \bar{r}}{(1 + (M - 1) \cdot \bar{r})},$$

whereby $\bar{r}$ represents the average correlation of the LOCs. Since we analyze only two LOCs (i.e., $M = 2$), their correlation (i.e., 0.631) is their average correlation and you obtain the following result for Cronbach's alpha:

$$\text{Cronbach's } \alpha = \frac{2 \times 0.631}{(1 + (2 - 1) \times 0.631)} = \frac{1.262}{1.631} = 0.774.$$

These internal consistency reliability results are well above the commonly recommended threshold of 0.7 (Hair et al., 2017).

In the next step, you can assess the discriminant validity between the LOCs *COMP* and *LIKE* using the HTMT criterion (Henseler et al., 2015). Exhibit 2.11 shows the HTMT values along with their 95% bias-corrected and accelerated (BCa) bootstrap confidence intervals.

The results indicate discriminant validity between *COMP* and *LIKE* since the HTMT value of 0.780 is below the (conservative) threshold value of 0.85. Furthermore, the corresponding bootstrap confidence interval does not include the value 1 (i.e., the HTMT value is significantly lower than 1). Also, we establish discriminant validity between LOCs and the reflectively measured construct *CUSL* as well as with the single item construct *CUSA*. At the same time, however, we cannot establish discriminant validity between *COMP* and *LIKE* and their HOC *REPU*. This result is expected, however, as the measurement model of the HOC repeats the indicators of its LOCs.

Since the results provide support for the measures' reliability and validity, you can now continue with the structural model evaluation. All structural model relationships are significant ($p \leq 0.01$). The antecedent constructs *QUAL* (0.444) and *PERF* (0.228) have the strongest effects on *REPU*, while *CSOR* (0.130) and *ATTR* (0.142) are less relevant (Exhibit 2.10). *REPU* itself has a strong effect on *CSUA* (0.527), which is in turn strongly related to *CUSL* (0.527). The direct relationship between *REPU* and *CUSL* is somewhat weaker (0.308). The R^2 value of all the endogenous latent variables (i.e., 0.715 for *REPU*, 0.277 for *CUSA*, and 0.543 for *CUSL*) is relatively high when taking the number of antecedent constructs into account.

Exhibit 2.11	**Discriminant Validity Assessment Using the HTMT Criterion**				
	COMP	*CUSA*	*CUSL*	*LIKE*	*REPU*
COMP					
CUSA	0.465 [0.351; 0.571]				
CUSL	0.532 [0.412; 0.643]	0.755 [0.688; 0.818]			
LIKE	0.780 [0.696; 0.858]	0.577 [0.487; 0.655]	0.737 [0.655; 0.816]		
REPU	1.100 [1.068; 1.143]	0.562 [0.472; 0.645]	0.685 [0.600; 0.766]	1.073 [1.048; 1.105]	

Note: The values in brackets represent the 95% bias-corrected and accelerated confidence interval of the HTMT values obtained by running the bootstrapping routine with 5,000 samples in SmartPLS.

Reflective-Formative HCM

As an alternative to the reflective-reflective HCM, one may consider that the two subdimensions *COMP* and *LIKE* form the general *REPU* construct. For this reason, we also illustrate the reflective-formative HCM application of the corporate reputation model. Exhibit 2.12 shows the PLS path model that results from these theoretical/conceptual considerations.

However, the model shown in Exhibit 2.12 represents an inadmissible reflective-formative HCM setup in PLS-SEM. Since the LOCs *COMP* and *LIKE* explain by (almost) 100% the variance of the HOC *REPU* ($R^2 \approx 1.00$), there is no remaining variance that its antecedent constructs *ATTR*, *CSOR*, *PERF*, and *QUAL* can explain. Hence the relationships of these four constructs with *REPU* in the PLS path model have path coefficient estimates close to zero (Exhibit 2.12). Concluding that these relationships are zero (and nonsignificant) would be premature as these estimates result from the setup of the reflective-formative HOC *REPU*.

A different picture emerges, however, when the total effects analysis of a collect-type HCM is specified. To put this analysis into practice, you will need to add further indirect relationships from *ATTR*, *CSOR*, *PERF*, and *QUAL* via *COMP* and *LIKE* to *REPU*, as shown in Exhibit 2.13. Next, run the PLS-SEM algorithm by clicking on **Calculate → PLS Algorithm** in the menu. Use the default settings except for the weighting scheme, which should be set to **Factor**. Using the results report that opens after running the PLS-SEM algorithm, you can now assess the reflective and formative measurement models. The results of this assessment are very similar to those reported in Hair et al. (2017). However, potential collinearity among the LOCs of the reflective-formative HCM demands close attention. To check for collinearity issues, go to **Quality Criteria → Collinearity Statistics (VIF)**. You find that the VIF values of *COMP* (2.719) and *LIKE* (2.296) are considerably below the threshold of 5, providing support that collinearity is not a critical issue. You also find that the VIF value of *QUAL* is relatively high (4.202) but still below the threshold. Next, go to **Final Results → Total Effects**, which opens the table shown in Exhibit 2.14. The column labeled *REPU* shows for each row the effects (1) of the LOCs *COMP* and *LIKE* on the HOC *REPU* and (2) of *REPU*'s antecedents in the PLS path model (i.e., *ATTR*, *CSOR*, *PERF*, and *QUAL*). You find that the two LOCs have highly similar effects on *REPU* (0.522 for *COMP* and 0.583 for *LIKE*) and thus have equal relevance for forming the HOC.

Inadmissible Reflective-Formative HCM Set-Up

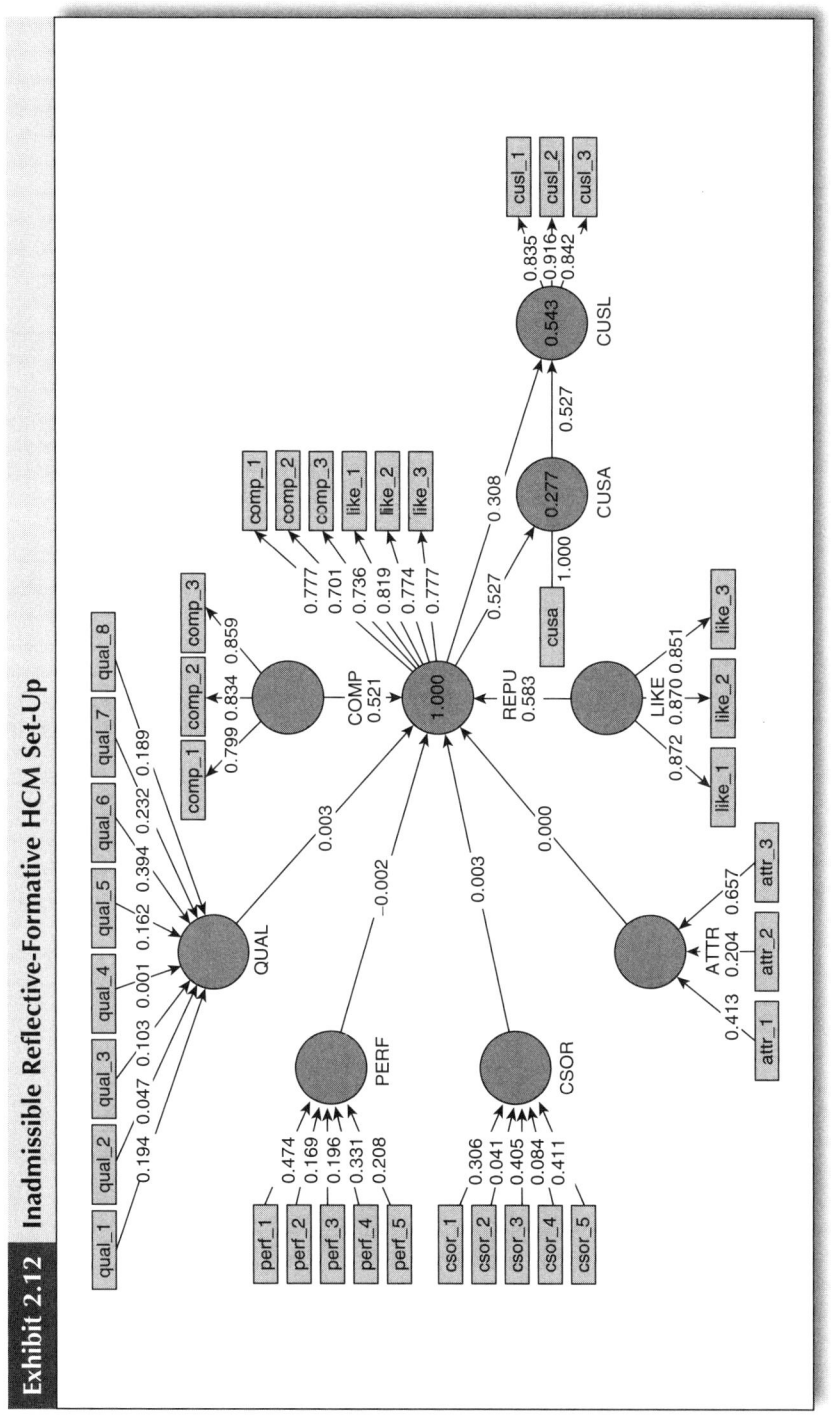

Exhibit 2.13 Total Effects Analysis of Corporate Reputation HCM

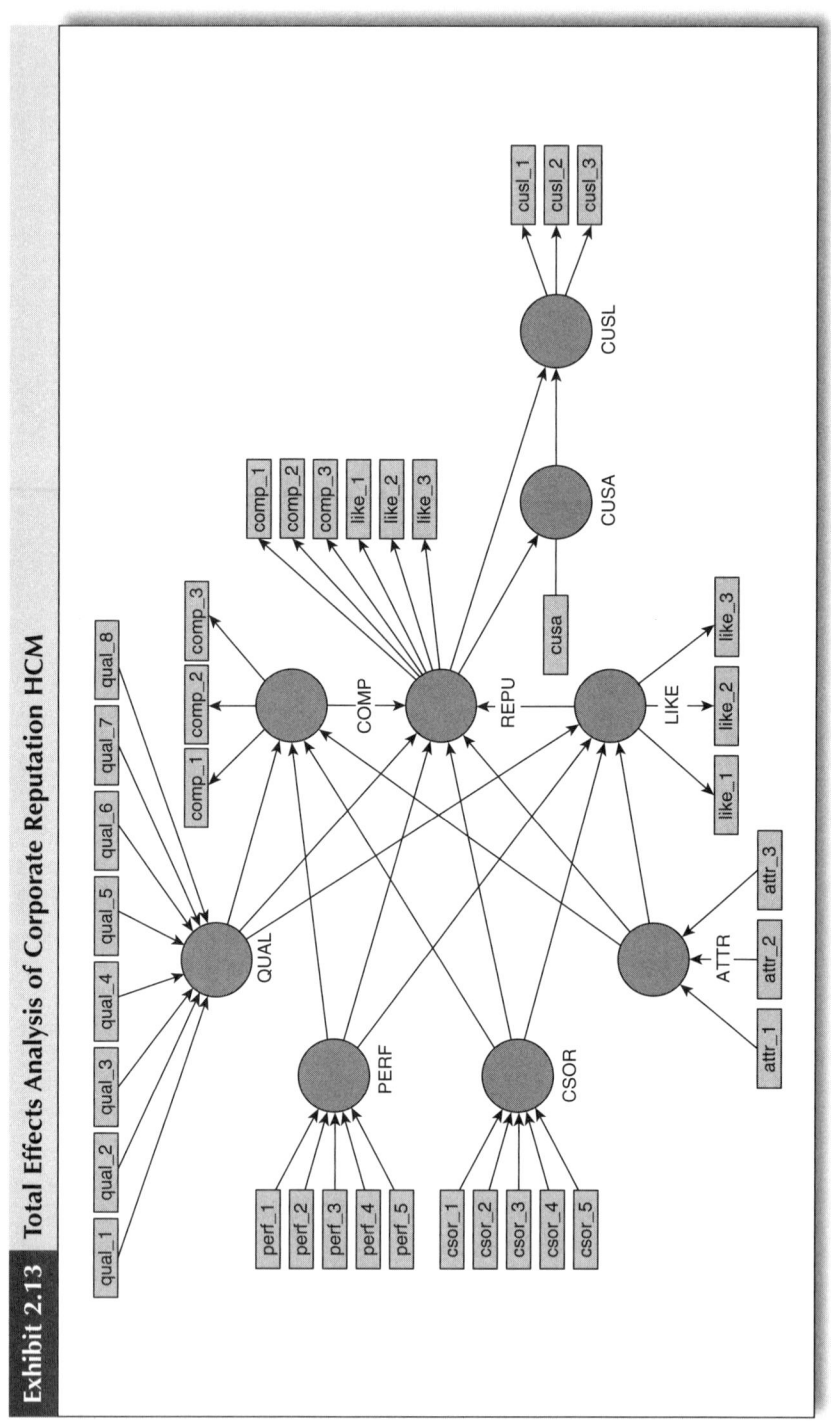

Exhibit 2.14	SmartPLS Results for the Total Effects

Total Effects

🔲 Matrix

	ATTR	COMP	CSOR	CUSA	CUSL	LIKE	PERF	QUAL	REPU
ATTR		0.087		0.075	0.083	0.166			0.142
COMP				0.275	0.306				0.522
CSOR		0.049		0.069	0.077	0.179			0.132
CUSA					0.527				
CUSL									
LIKE				0.307	0.342				0.583
PERF		0.305		0.119	0.133	0.118			0.227
QUAL		0.424		0.234	0.260	0.379			0.444
REPU				0.527	0.586				

Among *REPU*'s antecedents, *QUAL* has the strongest effect (0.444), followed by *PERF* (0.227), *ATTR* (0.142), and *CSOR* (0.132).

NONLINEAR RELATIONSHIPS

In PLS-SEM, relationships between constructs can take various forms. While **linear relationships** can be represented by straight lines (with positive or negative slopes) when plotting the latent variables' values in a scatterplot, **nonlinear relationships** include all associations that are not straight lines but curves. Nonlinear relationships occur quite frequently but are generally hard to determine a priori on theoretical grounds. Moreover, linear relationships often approximate nonlinear ones sufficiently well such that linear relationships appear satisfactory.

A typical example of a nonlinear relationship is the diffusion of new products as a function of time (e.g., Bass, 1969). Similarly, the relationship between marketing activities (e.g., advertising spending, promotional activities) and product sales usually follows a nonlinear relationship. Specifically, we would generally expect a positive relationship between marketing activities and sales. However, increasing marketing activities oftentimes involves a positive but diminishing effect on sales (Hay & Morris, 1991), meaning that after a point every additional unit of marketing activities contributes less to increasing sales. Similarly, another example is the widespread assumption of a positive linear relationship between satisfaction and loyalty, which is likely nonlinear (Eisenbeiss, Cornelißen, Backhaus, & Hoyer, 2014). With higher levels of satisfaction, at some point the strength of the positive effect on loyalty instead decreases; that is, the effect remains positive but reduces in size.

Establishing nonlinear effects requires careful theoretical reasoning. Also, preliminary analyses can support researchers in identifying nonlinear relationships by plotting values of two variables against each other. In a PLS-SEM analysis, this would involve comparing the scores of two latent variables in a scatterplot using spreadsheet software such as Microsoft Excel or statistical software such as IBM SPSS Statistics (see Sarstedt & Mooi, 2014).

The initial approach to handle nonlinear relationships draws on data transformations of one or both variables between which the relationship is expected or observed. A typical type is **log transformation**, which applies a base 10 logarithm to every observation. As logarithms cannot be calculated for negative values, which by definition occur in the standardized latent variable scores, researchers need to initially transform the indicators of the measurement models of the latent variable—rather than the latent variable scores themselves. If the transformation does not sufficiently linearize the relationship, we have to explicitly include the nonlinear relationship in the PLS path model.

To better grasp the concept of nonlinearity and how to examine nonlinear relationships in PLS path models (Wold, 1982), consider the following formal (**polynomial**) representation of a **nonlinear effect** (note that we excluded the constant term and the error term to simplify the illustration): $Y_2 = p_1 Y_1 + p_2 Y_1^2 + \ldots + p_N Y_1^N$. In this equation, Y_2 represents the endogenous latent variable and Y_1 the exogenous latent variable, while p_1 is the path coefficient of the linear relationship between Y_1 and Y_2. The nonlinear effect of the relationship is added by $p_N Y_1^N$ to the linear part $p_1 Y_1$, whereby N stands for the **polynomial degree**, which is the highest exponent occurring in the polynomial. If N equals one, we assume only a linear relationship (i.e., $Y_2 = p_1 Y_1$), while integer values for N larger than 1 determine different types of nonlinear functions. For example, for $N = 2$, the relationship follows a quadratic form: $Y_2 = p_1 Y_1 + p_2 Y_1^2$. For $N = 3$, the relationship is cubic: $Y_2 = p_1 Y_1 + p_2 Y_1^2 + p_3 Y_1^3$. The polynomial degree determines the form of the nonlinear function, which has approximately $N-1$ turning points where the slope of the curve changes signs. For example, when $N = 2$, the resulting quadratic function has one turning point (e.g., the slope's negative sign changes to a positive one as shown in the top left box of Exhibit 2.15); when $N = 3$, the ensuing cubic function shows two turning points (e.g., with sign changes of the slope from a negative to a positive, and again to a negative one as shown in the top left box of Exhibit 2.16).

Exhibit 2.15 shows alternative forms of **quadratic effects** with varying values of p_1 and p_2. The p_1 coefficient indicates whether the overall linear relationship between Y_1 and Y_2 is positive or negative. The nonlinear effect p_2 determines the direction of the curvature. For example, the dotted lines in the two top graphs of Exhibit 2.15 indicate that the overall relationship (i.e., the **linear effect**) between Y_1 and Y_2 is positive. Adding a positive quadratic term creates a convex or u-shaped curve (solid line in the top left box in Exhibit 2.15). Conversely, adding a negative quadratic term creates a concave or an inverse u-shaped curve (solid line in the top right box in Exhibit 2.15). The two bottom graphs in Exhibit 2.15 represent the same situations but with a negative linear effect between Y_1 and Y_2.

Exhibit 2.15	**Examples of Quadratic Effects**

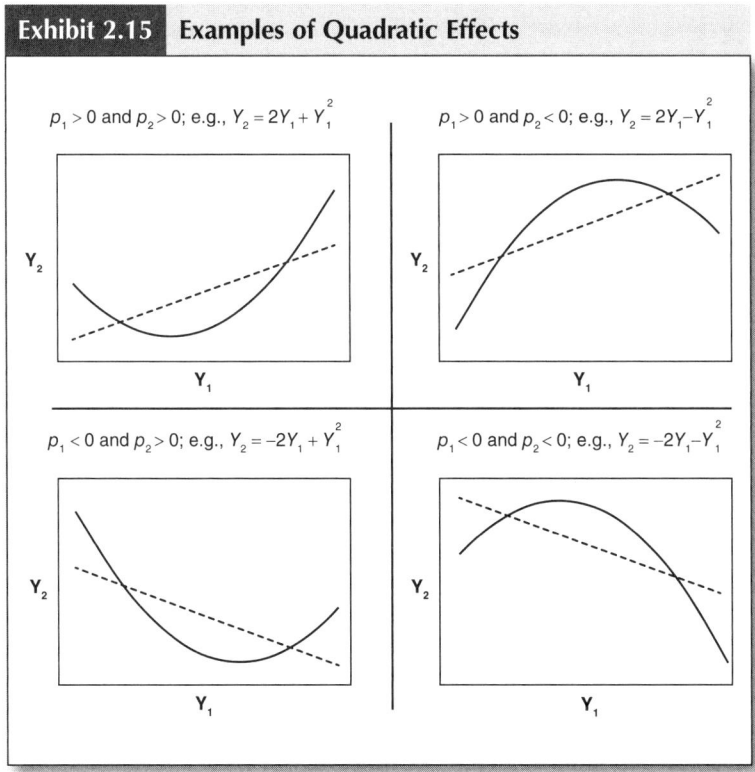

Exhibit 2.16 shows forms of **cubic effects** with varying values of p_1, p_2, and p_3. Compared to the quadratic effect models in Exhibit 2.15, the curves representing the cubic effects have two turning points (i.e., sign changes of their slopes). For the example in the top left box of Exhibit 2.16 with positive p_1 and p_2 values but a negative one for p_3, the cubic function is convex at first but changes into a concave function.

Exhibits 2.15 and 2.16 display typical shapes of nonlinear functions for a wide range of Y_1 and Y_2 values. However, in standard regression analyses, values are frequently limited to a smaller range that includes only certain areas of the overall function. For example, consider a cubic effect as shown in the left box of Exhibit 2.16 and let only the middle part of the function fall into the relevant range of values for the analysis (e.g., approximately from the function's first turning point on the left-hand side to second turning point on the right-hand side). Then, instead of assuming a cubic function, one would focus only on a slightly increasing function, which almost is linear (with a positive slope). Besides considering the different possible nonlinear functions, the relevant value ranges pose an additional difficulty in making a priori assumptions about the relevant shape of the relationship between two latent variables. In PLS-SEM, however, latent variable scores are standardized, which complicates censoring the value range.

In general, hypothesizing quadratic effects already represents a very challenging task. However, theorizing the existence of a cubic effect is even more difficult. For this reason, considering simpler

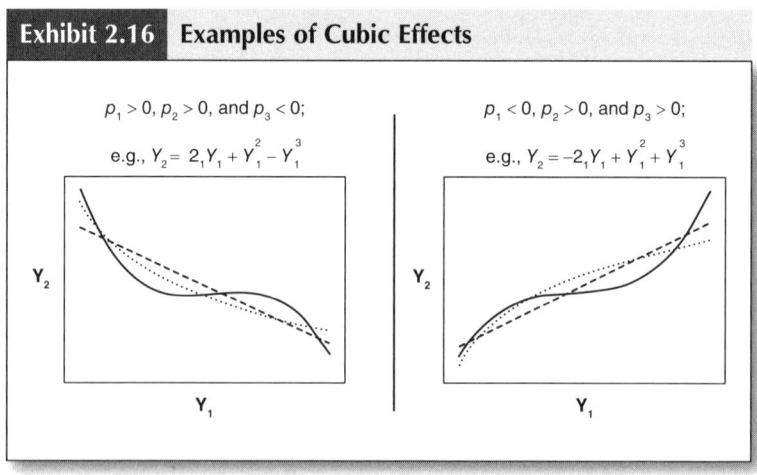

Exhibit 2.16	**Examples of Cubic Effects**

$p_1 > 0, p_2 > 0,$ and $p_3 < 0;$

e.g., $Y_2 = 2_1Y_1 + Y_1^2 - Y_1^3$

$p_1 < 0, p_2 > 0,$ and $p_3 > 0;$

e.g., $Y_2 = -2_1Y_1 + Y_1^2 + Y_1^3$

Note: dashed line = linear; dotted line = quadratic; solid line = cubic.

quadratic effects is typically the primary focus when modeling and testing nonlinear effects unless theory provides compelling evidence for more complex nonlinear relationships. In social science models, quadratic effects occur more commonly than other types of nonlinear effects. Therefore, we focus on this type of nonlinear effect in the remainder of this chapter.

Modeling Quadratic Effects in PLS-SEM

To understand how quadratic effects are implemented in PLS path models, consider the quadratic equation $Y_2 = p_1 Y_1 + p_2 Y_1^2$. This model is specified in a linear way as the path coefficients p_1 and p_2 that link Y_1 and Y_1^2 with Y_2 are linear (e.g., Sarstedt & Mooi, 2014). However, the nonlinear (i.e., quadratic) effect is introduced with the Y_1^2 variable, which entails an interaction of Y_1 with itself (i.e., $Y_1 \cdot Y_1$). More precisely, a quadratic effect can be conceived as a special case of a moderation model (Hair et al., 2017) in which Y_1 self-moderates the relationship between Y_1 and Y_2 (Exhibit 2.17). That is, the linear relationship between Y_1 and Y_2 changes in size depending on the values of Y_1.

In line with Rigdon, Ringle, and Sarstedt (2010), similar to the interaction effects in standard moderator analysis, we can model self-moderation by using an **interaction term**. But different from a standard moderator analysis, where the interaction term represents the interaction between Y_1 and some other moderator variable, in self-interaction the quadratic term embodies the interaction of Y_1 with itself—as expressed in the following equation: $Y_2 = (p_1 + p_2 Y_1) \cdot Y_1 = p_1 Y_1 + p_2 Y_1^2$. If the quadratic effect is positive, the strength of Y_1's effect on Y_2 increases for higher values of Y_1. Conversely, if the quadratic effect is negative, higher values in Y_1 imply a lower effect of Y_1 on Y_2. Exhibit 2.18 illustrates the concept of self-interaction with a quadratic term (i.e., Y_1^2 with the relationship p_2 to Y_2) as an additional latent variable in the PLS path model, covering the product of Y_1 with itself, in addition to the

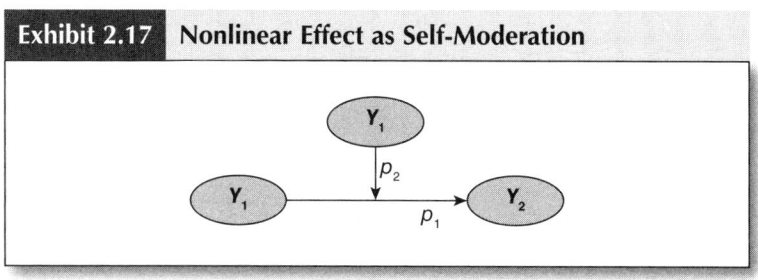

Exhibit 2.17   **Nonlinear Effect as Self-Moderation**

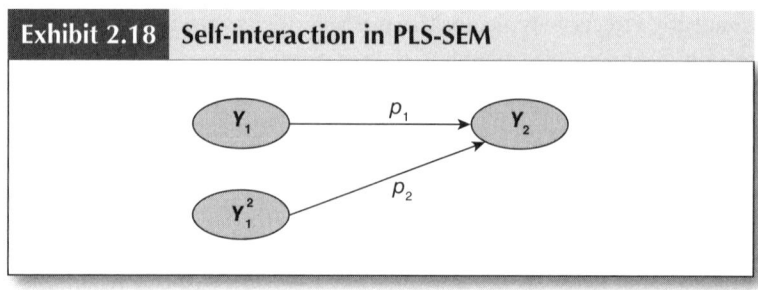

Exhibit 2.18 | **Self-interaction in PLS-SEM**

linear p_1 relationship from Y_1 to Y_2. Similarly, in the case of a cubic relationship, one would additionally include the construct Y_1^3 with the relationship p_3 to Y_2.

To create the quadratic term Y_1^2 in the PLS path model, the same principles apply as in a standard moderator analysis. Accordingly, researchers can draw on the following approaches (Hair et al., 2017): the product indicator approach, the orthogonalizing approach, and the two-stage approach.

The **product indicator approach** is the standard approach for creating the interaction term in regression-based analyses and also features prominently in PLS-SEM. The product indicator approach to establish the quadratic effect in PLS-SEM involves multiplying each standardized indicator of the exogenous latent variable Y_1 with itself and all other indicators in the same measurement model of the latent variable. These so-called **product indicators** become the indicators of the quadratic term. The first element in Exhibit 2.19 illustrates the product indicator approach for our previous model, assuming that Y_1 is measured with two indicators (x_1 and x_2). Correspondingly, the quadratic term has three product indicators, namely $x_1 \cdot x_1$, $x_1 \cdot x_2$, and $x_2 \cdot x_2$. As in standard moderator analysis, this approach is applicable only when the exogenous construct Y_1 is measured reflectively. Furthermore, as the approach requires the indicators of the exogenous construct to be reused in the measurement model of the quadratic term, the product indicator approach inevitably introduces collinearity in the path model, but this can be dealt with easily by using the orthogonalizing approach.

The **orthogonalizing approach** is an extension of the product indicator approach and provides a solution to the collinearity issue of the product indicator approach. To begin with, as in the product indicator approach, the orthogonalizing approach creates all product indicators of the quadratic term, which are $x_1 \cdot x_1$, $x_1 \cdot x_2$, and $x_2 \cdot x_2$ in our example. Then, it regresses each product indicator on all indicators of the exogenous construct Y_1. In each regression model, a product indicator (e.g., $x_1 \cdot x_1$) represents the dependent variable, while all indicators of the exogenous construct (here, x_1 and x_2) act as independent variables:

$$x_1 \cdot x_1 = b_1 \cdot x_1 + b_2 \cdot x_2 + e_{11}$$
$$x_1 \cdot x_2 = b_3 \cdot x_1 + b_4 \cdot x_2 + e_{12}$$
$$x_2 \cdot x_2 = b_5 \cdot x_1 + b_6 \cdot x_2 + e_{22}$$

When examining the results, we are not interested in the regression coefficients b_1 to b_6 but in the standardized residual terms e_{11}, e_{12}, and e_{22},

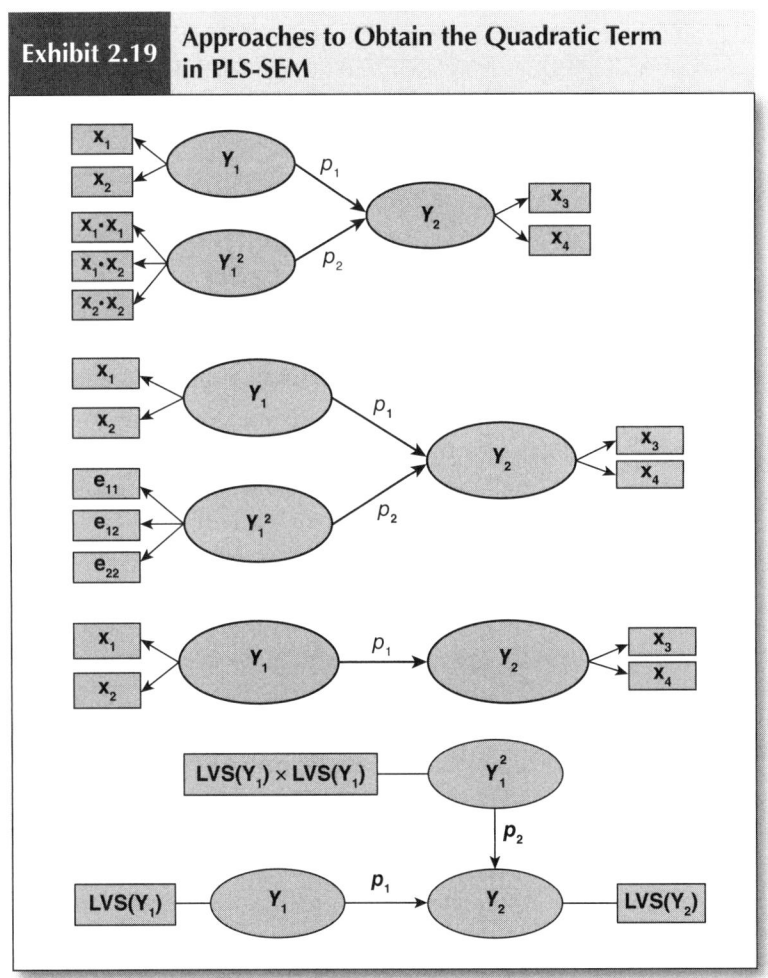

| Exhibit 2.19 | Approaches to Obtain the Quadratic Term in PLS-SEM |

which become the indicators of the quadratic term (Exhibit 2.19). This analysis ensures that the indicators of the quadratic term do not share any variance with any of the exogenous construct Y_1's indicators. In other words, the quadratic term is orthogonal to Y_1, precluding any collinearity issues among the constructs used for establishing the linear and the quadratic effects. As the orthogonalizing approach rests on the product indicator approach, it is also applicable only when the exogenous construct Y_1 is measured reflectively.

The two-stage approach (nonlinear effects) incorporates the following procedure: (1) Estimate the model without the interaction term and obtain the latent variable scores, and (2) use the latent variable scores as indicators of the latent variables in the nonlinear model (LVS in Exhibit 2.19). Here, the element-wise product of the standardized latent variable scores of the exogenous variable Y_1 with itself serves as an indicator of the quadratic term. The advantage of this approach is that it can be applied to both reflectively and formatively measured exogenous latent variables.

To specify which approach should be preferred to create the quadratic term, we draw on Henseler and Chin (2010), who ran an extensive simulation study on moderation analysis. They compared the approaches in terms of their statistical power, point estimation accuracy, and prediction accuracy. We also rely on the recommendations by Hair et al. (2017, Chapter 7). In general, the orthogonalizing approach proves valuable when the objective of the analysis is to minimize the estimation bias of the nonlinear effect or to maximize prediction. But when the objective is to identify the statistical significance of the nonlinear effect, the two-stage approach is recommended. Overall, the two-stage approach is the most versatile approach as it can be used regardless of whether the exogenous construct is measured reflectively or formatively. Therefore, the two-stage approach is generally preferred.

Finally, and similar to the interaction in the moderator analysis (Hair et al., 2017), it is important to note that the estimated values of p_1 and p_2 represent the strength of the relationship between Y_1 and Y_2 when Y_1 has a value of zero. In many model setups, however, zero is not a number on the scale of Y_1. If this is the case, the interpretation of the quadratic effect becomes problematic. This is the reason why the indicators of the independent variable Y_1 need to be mean-centered or standardized. Since PLS-SEM uses standardized data, we focus on this option. Standardization is achieved by subtracting the variable's mean from each observation and dividing the result by the variable's standard error (Sarstedt & Mooi, 2014). After standardization, Y_1's average value represents the point of reference in the quadratic model, which facilitates interpretation of the effects.

Evaluation of Nonlinear Effects

Measurement and structural model evaluation criteria also apply to nonlinear models (for more details, see Hair et al., 2017). The reflectively measured exogenous latent variable Y_1 must meet all relevant criteria in terms of internal consistency reliability, convergent validity, and discriminant validity. Similarly, all formative measurement model assessment criteria must be considered respectively. For the quadratic term, however, there is no such requirement as this construct serves only as an auxiliary measurement, designed to model the quadratic effect. Therefore, similar to the interaction term in standard moderator analysis, the quadratic term's measurement model does not have to be assessed (Hair et al., 2017). In the two-stage approach, the evaluation criteria apply only to the first stage but not to the second stage since the latter involves only single item constructs.

Following the analysis of the measurement models, the next step requires analyzing the significance of the quadratic effect. The test procedure involves generating the distribution of the parameter by using the bootstrapping procedure (e.g., with 5,000 samples using the no sign changes option; for more detail, see Hair et al., 2017). If the 95% confidence interval of the nonlinear effect does not include the value zero, the nonlinear effect is significantly different from zero at a 5% significance level. In that case, we conclude that the variable of interest has a significant nonlinear effect. Alternatively, we can examine the p value or the corresponding t value to assess whether the nonlinear effect is significant.

When analyzing the significance of effects, here, too, we have to keep in mind that significance does not imply relevance. Particularly with larger sample sizes, very small effects can become significant. But this does not imply that these effects are relevant. Therefore, the next step is to assess the strength of the nonlinear effect by means of the f^2 effect size. The f^2 effect size assesses the change in the R^2 value of the endogenous latent variable Y_2 when the exogenous quadratic term Y_1^2 is omitted from the model. The f^2 effect size value is computed as follows:

$$f^2 = \frac{R^2_{\text{model with quadratic effect}} - R^2_{\text{model without quadratic effect}}}{1 - R^2_{\text{model with quadratic effect}}}$$

Consequently, the f^2 effect size indicates how much the quadratic effect contributes to the explanation of the endogenous latent variable. General guidelines for assessing f^2 suggest that values of 0.02, 0.15, and 0.35 represent small, medium, and large effect sizes, respectively (Cohen, 1988). However, when assessing arguments that apply to moderation effects, Aguinis, Beaty, Boik, and Pierce (2005) point out that the average effect size in tests of moderation is only 0.009. Against that background, Kenny (2015) proposes that 0.005, 0.01, and 0.025 constitute more realistic standards for small, medium, and large effect sizes, respectively. But he also points out that even these values are optimistic given Aguinis et al.'s (2005) review.

Note again that linear effects more often than not offer a reasonable approximation of quadratic effects. Additionally, standard considerations concerning statistical power have to be adhered to (see Hair et al., 2017). Therefore, unless deviations of normality are extreme, researchers should carefully examine whether the inclusion of a quadratic term adds to the analysis of linear effects.

Results Interpretation

For the interpretation of quadratic effects, consider the estimation results of the following linear model with (unstandardized) coefficients: $Y_2 = 0.2 \cdot Y_1$. The estimate of $p_1 = 0.2$ indicates the effect of a one-unit change in Y_1 on Y_2. More precisely, every additional unit of Y_1 increases Y_2 by 0.2 units. Now, if we additionally consider a quadratic effect, the formula could change into $Y_2 = 0.6 \cdot Y_1 - 0.15 \cdot Y_1^2$. The coefficient p_1 for the linear Y_1 term changed from 0.2 in the simple linear model to 0.6 in the quadratic model. However, this result does not mean that the linear effect tripled for a unit change in Y_1 since the two estimates of p_1 are not directly comparable. More specifically, the estimate of $p_1 = 0.6$ in the quadratic model does not imply that an increase of one unit of Y_1 changes Y_2 by 0.6, because another Y_1 term exists in the equation (i.e., the quadratic term Y_1^2). Both terms' effects must be considered simultaneously in order to correctly infer the relationship between Y_1 on Y_2. For example, when inserting the values of 0, 1, and 2 for Y_1 into the quadratic model (i.e., $Y_2 = 0.6 \cdot Y_1 - 0.15 \cdot Y_1^2$), we obtain the following results:

$$Y_1 = 0 \implies Y_2 = 0.6 \cdot 0 - 0.15 \cdot 0 = 0$$
$$Y_1 = 1 \implies Y_2 = 0.6 \cdot 1 - 0.15 \cdot 1 = 0.45$$
$$Y_1 = 2 \implies Y_2 = 0.6 \cdot 2 - 0.15 \cdot 4 = 0.60$$

As can be seen, a one-unit change of Y_1 from 0 to 1 entails a change of Y_2 by 0.45 units while a one-unit change from 1 to 2 entails a considerably smaller change of Y_2 by 0.15 units. Hence, in contrast to the linear relationship, the slope and thus the effect of a one-unit change depends on the level of Y_1. For this reason, plotting the quadratic function as shown in Exhibit 2.15 is very useful for presentation and interpretation of the results.

The interpretation of the quadratic effect is similar when considering standardized data and parameter estimates, as is common in PLS-SEM analyses. In this situation, the coefficients indicate the effect of changes in a latent variable's values with regard to standard deviations, rather than its values as such. More precisely, considering the quadratic equation above, zero represents Y_1's average value as the point of reference for the analysis. If the level of Y_1 is increased (or decreased) by one standard deviation unit, Y_1's linear effect on Y_2 changes by the size of the quadratic term p_2. For example, consider the previous example where the linear effect p_1 equals 0.60 and the quadratic effect p_2 has a value of -0.15. If the value of the exogenous construct Y_1 increases by one standard deviation unit, one would expect the linear relationship between Y_1 and Y_2 to decrease by 0.15 units, so the effect at the higher level of Y_1 equals $0.60 - 0.15 = 0.45$. These are ceteris paribus considerations, which means that the aforementioned changes are anticipated while everything else in the PLS path model remains constant.

Exhibit 2.20 summarizes the rules of thumb for analyzing nonlinear effects in PLS-SEM.

Case Study Illustration

To illustrate the estimation of nonlinear effects, consider the extended corporate reputation model again. As previously indicated,

Exhibit 2.20	**Rules of Thumb for Analyzing Nonlinear Effects in PLS-SEM**

- Establish your expectations of nonlinear relationships a priori based on theoretical considerations or prior analysis of the relationships between latent variables by means of scatterplots. Keep in mind that linear effects generally offer reasonable approximations of nonlinear effects.

- For the creation of the nonlinear term, use the two-stage approach when the exogenous construct is measured formatively or when the aim is to disclose the significance of a presumed nonlinear effect. Alternatively, use the orthogonalizing approach, especially when the aim is to minimize estimation bias of the quadratic effect or to maximize prediction. Independent from these aspects, the two-stage approach is very versatile and should generally be given preference for creating the interaction term.

- Create the quadratic term or any other nonlinear effect based on standardized data.

- The exogenous and endogenous variables must be assessed for reliability and validity following the standard evaluation procedures for reflective and formative measures. This does not hold for the nonlinear term, however, which relies on an auxiliary measurement model generated by reusing indicators of the exogenous construct.

- Evaluate whether the nonlinear effects are significant by using the 95% bootstrap confidence intervals. For their significant nonlinear effects, assess their relevance based on their f^2 effects size. Outcomes higher than 0.005, 0.01, and 0.025 constitute small, medium, and large f^2 effect sizes.

theory suggests that the effect of customer satisfaction (*CUSA*) on customer loyalty (*CUSL*) is nonlinear (e.g., Eisenbeiss et al., 2014). For this reason, we introduce the quadratic effect of *CUSA* into the model assuming that a positive effect of *CUSA* on *CUSL* diminishes as the level of *CUSA* increases. The SmartPLS software offers an option to automatically include a quadratic term. Right-click in the target construct *CUSL* and choose the option **Add Quadratic Effect** (Exhibit 2.21).

In the screen that follows, specify *CUSA* as the **Independent Variable**. Under **Calculation Method**, you can choose between the product indicator, the orthogonalizing, and the two-stage approach for creating the quadratic term. In this case study, our primary concern is to disclose the significance of the nonlinear effect (which is

Exhibit 2.21	Add Quadratic Effect Menu Option

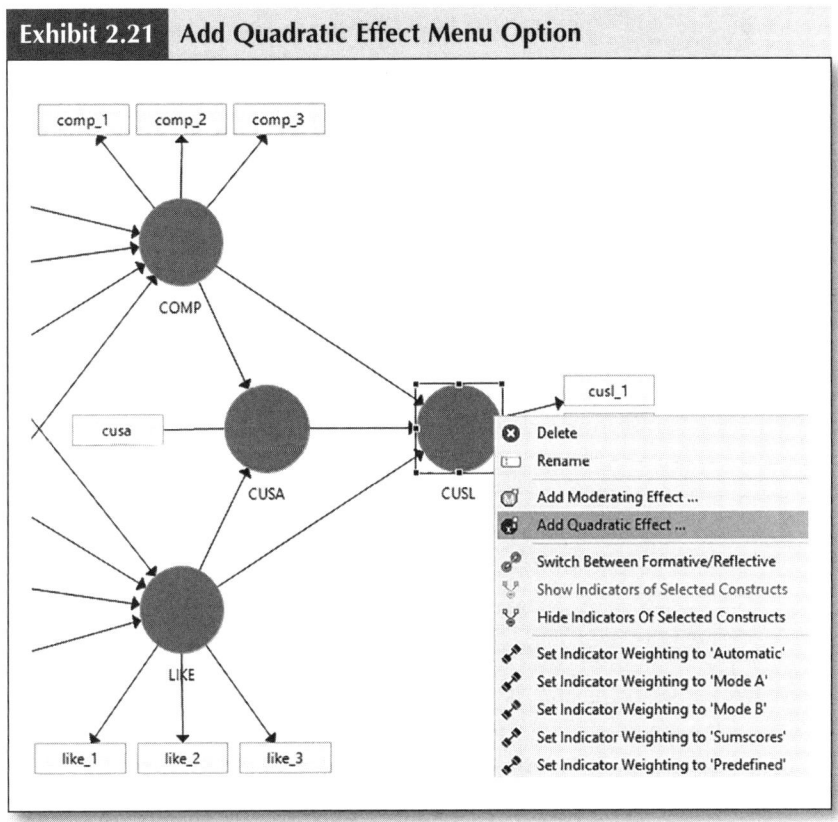

usually the case in PLS-SEM applications), so the two-stage approach is appropriate. Choose the **Two-Stage** option as well as **Standardized** and **Automatic** under **Advanced Settings** (Exhibit 2.22).

When you click **OK,** SmartPLS will include the nonlinear term labeled *Quadratic Effect 1* in the modeling window. If you like, you can right-click on this construct and select the **Rename** option to choose a different name (e.g., $CUSA^2$). Its different color also indicates that this construct is a quadratic term. Right-click on the quadratic term and choose the menu option **Show Indicators of Selected Constructs.** The indicator *CUSA* generated in the second stage of the two-stage approach will then appear in the modeling window. You can now proceed with the analysis by running the PLS-SEM algorithm (using the path weighting scheme and mean value replacement for missing values). Exhibit 2.23 shows the results in the modeling window. The following equation illustrates the results of the relationship between *CUSA* and *CUSL*: $CUSL = 0.467 \cdot CUSA - 0.046\ CUSA^2$.

Exhibit 2.22	Quadratic Effect Dialogue Box in SmartPLS

Quadratic Effect

— **Basic Settings** ─────────────────────

Dependent Variable   CUSL

Independent Variable   CUSA   ⌄

Calculation Method   ○ Product Indicator
                       ◉ Two Stage
                       ○ Orthogonalization

— **Advanced Settings** ─────────────────

Product Term Generation   ○ Unstandardized
                             ○ Mean Centered
                             ◉ Standardized

Weighing Mode   ◉ Automatic
                             ○ Mode A
                             ○ Mode B
                             ○ Sumscores
                             ○ Pre Defined

Since *CUSA* is a single item construct, you cannot apply measurement model evaluation criteria and, thus, directly focus on the significance of the quadratic term $CUSA^2$, which has a relationship of −0.046 with *CUSA*. To assess whether this quadratic term is significant, run the bootstrapping procedure by going to **Calculate** → **Bootstrapping** in the SmartPLS menu or going to the **Modeling** window and clicking on the **Calculate** icon, followed by **Bootstrapping** (note that you first may need to go back to the **Modeling** window before the **Calculate** icon appears). You retain all settings for missing value treatment and the PLS-SEM algorithm as in the initial model estimation and select the **No Sign Changes** option, **5,000** bootstrap samples, and the **Basic Bootstrapping** option. In the advanced settings, choose **Bias-Corrected and Accelerated (BCa) Bootstrap, two-tailed** testing, and a significance level of **0.05** (Exhibit 2.24). Next, click on **Start Calculation**. After running the procedure, SmartPLS shows the bootstrapping results for the measurement models and structural model in the modeling window. Using the **Calculation Results** box at the bottom left of the screen, you can choose whether SmartPLS should display *t* values or *p* values in the modeling window.

Exhibit 2.23 Quadratic Effect Results of *CUSA* in SmartPLS

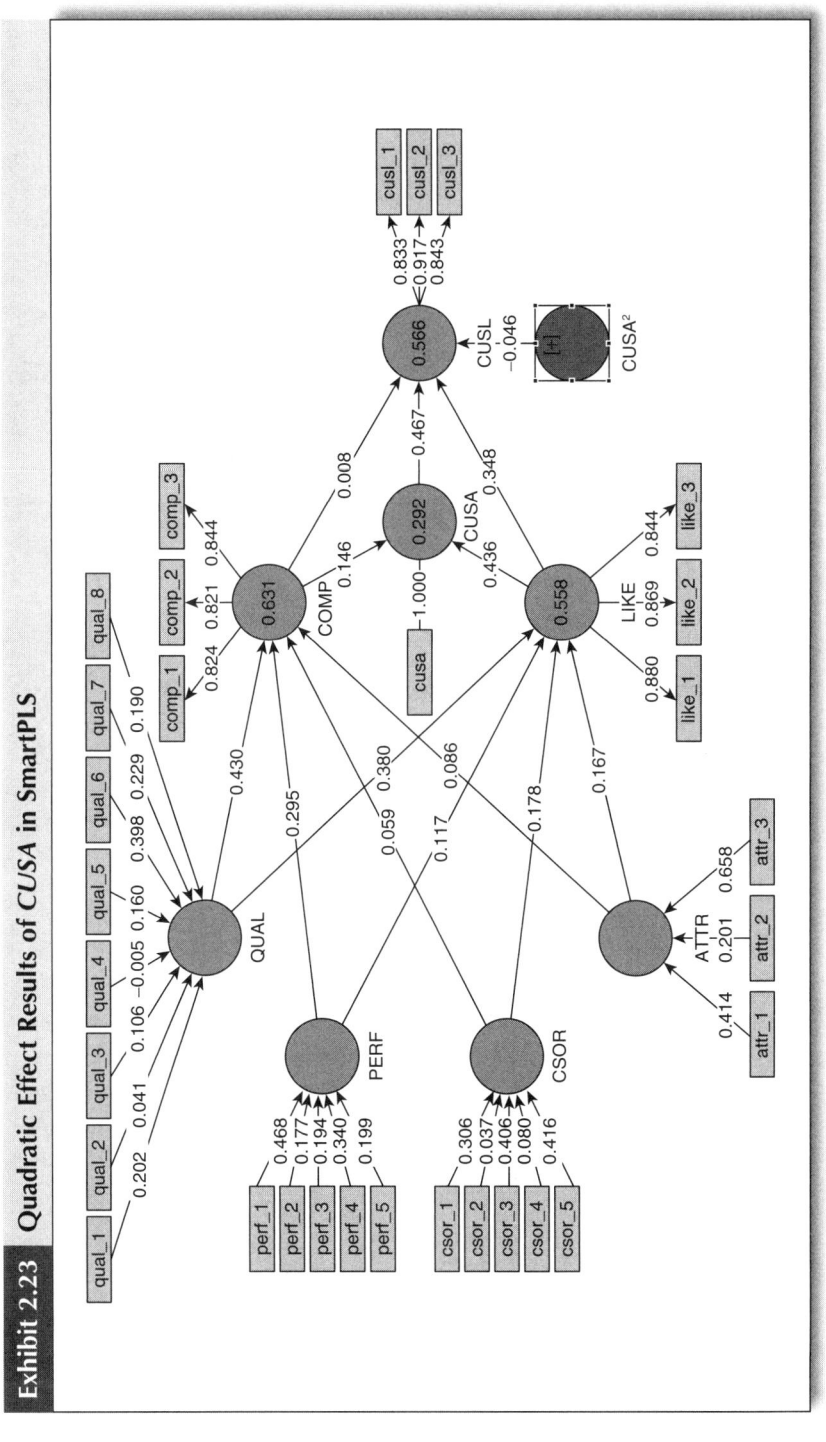

Exhibit 2.24	Bootstrapping Dialogue in SmartPLS

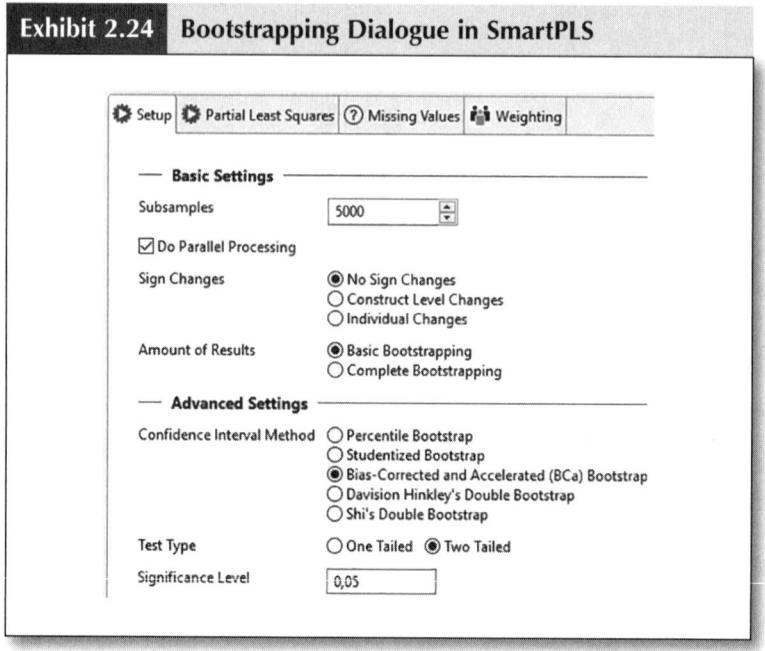

We obtain a p value of 0.055 and can conclude that the quadratic term $CUSA^2$ is not significant at a 5% significance level. Note that the results presented here will likely differ from your results and will change again when rerunning bootstrapping as the procedure builds on randomly drawn samples. By going to the bootstrapping results report, you can access a more detailed overview of the results. The table under **Final Results → Path Coefficients** provides an overview of results, including standard errors, bootstrap mean values, t values, and p values. The bootstrapping results report also provides bootstrap confidence intervals. Clicking on the **Confidence Intervals Bias Corrected** tab in the bootstrapping results report shows the confidence intervals generated from the BCa method (Exhibit 2.25). For the $CUSA^2 \rightarrow CUSL$ relationship, you find that zero falls into the 95% bias corrected confidence interval with a lower bound of –0.093 and an upper bound of 0.001. Hence, also based on the 95% bias corrected confidence interval, you can conclude that $CUSA$ does not have a significant quadratic effect on $CUSL$.

At this point, a quadratic relationship would be rejected and the significant linear relationship between $CUSA$ and $CUSL$ would be

Exhibit 2.25	**Bias-Corrected Bootstrap Confidence Intervals**

Extended model.splsm   Bootstrapping (Run No. 1)

Path Coefficients

Mean, STDEV, T-Values, P-Values | Confidence Intervals | Confidence Intervals Bias Corrected | Samples

	Original Sample (O)	Sample Mean (M)	Bias	2.5%	97.5%
ATTR -> COMP	0.086	0.085	-0.001	-0.015	0.195
ATTR -> LIKE	0.167	0.163	-0.004	0.043	0.289
COMP -> CUSA	0.146	0.145	-0.001	0.012	0.279
COMP -> CUSL	0.008	0.008	0.000	-0.094	0.116
CSOR -> COMP	0.059	0.062	0.004	-0.047	0.169
CSOR -> LIKE	0.178	0.179	0.000	0.063	0.286
CUSA -> CUSL	0.467	0.467	-0.000	0.381	0.554
CUSA^2 -> CUSL	-0.046	-0.048	-0.002	-0.093	0.001
LIKE -> CUSA	0.436	0.437	0.002	0.318	0.546
LIKE -> CUSL	0.348	0.348	0.001	0.239	0.453
PERF -> COMP	0.295	0.300	0.005	0.157	0.417
PERF -> LIKE	0.117	0.120	0.003	-0.023	0.252
QUAL -> COMP	0.430	0.428	-0.001	0.298	0.563
QUAL -> LIKE	0.380	0.386	0.006	0.249	0.505

retained. For illustrative purposes only, however, we continue the analysis even though the quadratic term is not significant.

The interpretation of *CUSA*'s quadratic effect uses the significant linear effect from *CUSA* to *CUSL* with a value of 0.467 as point of reference (Exhibit 2.25). For an increase of *CUSA* by one standard deviation unit, you add the quadratic effect to the linear one. Since the quadratic term is negative, one expects the relationship between *CUSA* and *CUSL* to become weaker (i.e., $0.467 - 0.046$) when *CUSA* increases by one standard deviation unit. On the contrary, when *CUSA* decreases by one standard deviation unit, one expects the relationship between *CUSA* and *CUSL* to become stronger (i.e., $0.467 + 0.046$). Thereby, you obtain an expected quadratic effect between *CUSA* and *CUSL* similar to the one shown in Exhibit 2.26.

However, because bootstrapping builds on random data generation, your results may slightly differ from those presented here. Thus, since the significance testing results generated in this illustration are very close to the 5% probability of error that we have specified as

Exhibit 2.26	Quadratic Relationship Between Customer Satisfaction and Loyalty

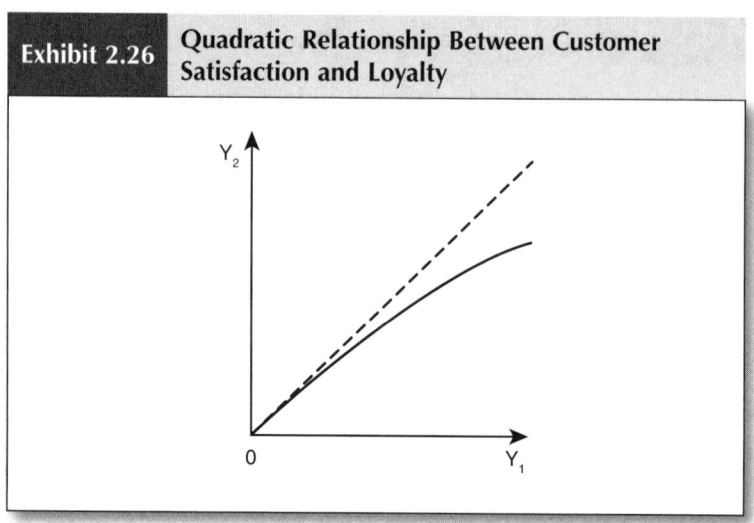

the critical level, results that you produce may reveal that the quadratic effect of *CUSA* and *CUSL* is significant. Hence, we advise repeating bootstrapping analyses to gain confidence in the results that are produced. Additional analysis that assists in understanding whether a nonlinear relationship is meaningful or not concerns examining the relevance of the quadratic effect. The small size of the coefficient (i.e., −0.046) in this example already indicates—whether significant or not—that the quadratic term is not particularly relevant. Hence, the final step of the results assessment addresses the quadratic term's f^2 effect size. By going to **Quality Criteria → f Square** in the SmartPLS algorithm results report, you find that the quadratic term's f^2 effect size has a value of 0.01. This value falls below the lower limit of 0.02 that, according to Cohen (1988), would at a minimum represent a small effect size. The more liberal interpretation by Kenny (2015), who proposes that values larger than 0.005, 0.01, and 0.025 establish small, medium, and large effect sizes, suggests a small to medium effect size.

To summarize, the nonsignificance of the quadratic effect and its low f^2 effect size suggest that you should consider a linear relationship instead of a nonlinear one. While both approaches entail very similar results (Exhibit 2.26), the linear relationship is less complex when it comes to making a priori assumptions about the relationship, the formal representation, the modeling of the relationship in PLS-SEM, and the interpretation of results.

SUMMARY

- Understand the usefulness of hierarchical component models (HCMs). An HCM embraces a more general construct (i.e., the HOC), measured at a higher level of abstraction, while simultaneously including several subcomponents (i.e., the LOCs), which cover more concrete traits of this construct. HCMs enable reducing the number of structural model relationships, making the PLS path model more parsimonious, while increasing the bandwidth of content covered by the respective constructs.

- Appreciate the different types of HCMs and understand how to use them in PLS-SEM. The establishment of HCMs builds on carefully established theoretical/conceptual considerations. On these grounds, researchers choose from four major HCM types. Each of these HCM types depicts the specific relationship between the HOC and the LOCs as well as the measurement model used to operationalize the constructs on the lower-order level: reflective-reflective, reflective-formative, formative-reflective, and formative-formative. Generally, the HOC of reflective-reflective and formative-reflective HCMs represents a more general construct that—similar to reflective measurement models—simultaneously explains all the underlying LOCs. Conversely, the HOC is formed by the LOCs in reflective-formative and formative-formative HCMs, which is similar to formative measurement models. The repeated indicators approach, the total effects analysis of a collect-type HCM, and the two-stage approach allow modeling and estimating HCMs in PLS-SEM. When specifying and estimating HCMs in PLS-SEM, researchers need to consider further aspects, which relate to the number of indicators per LOC, the PLS-SEM algorithm weighting scheme, and the use of Mode A and Mode B weighting.

- Comprehend how to run an HCM analysis using the SmartPLS software and how to interpret the results. Researchers can use SmartPLS to model any of the four HCM types introduced in this chapter. When analyzing the results of an HCM estimation, researchers need to carefully evaluate not only the measurement models of the

LOCs but also the measurement model of the HOC. Different from other constructs in the PLS path model, the assessment of the HOC is not concerned with the relationships between the HOC and its indicator variables but the relationships between the HOC and its LOCs. While these relationships are mapped as path coefficients in a PLS-SEM analysis, from a modeling perspective, they correspond to loadings (in case of reflective-reflective and formative-reflective HCMs) or weights (in case of reflective-formative or formative-formative HCMs) and need to be interpreted as such.

- **Understand the basic concepts of nonlinear analysis in a PLS-SEM context.** Nonlinear effects occur when the relationship between two constructs does not follow a straight line but a curve when plotting the latent variables' values in a scatterplot. When the relationship between two constructs is nonlinear, the size of the effect between two constructs depends on the values and the magnitude of change in the exogenous construct. When analyzing nonlinear effects, researchers have to make an assumption regarding the nature of the effect. While numerous different effect types are possible, quadratic effects are most common.

- **Appreciate how to use the SmartPLS software to estimate quadratic effects.** Analyzing a quadratic effect in PLS-SEM requires researchers to include an interaction term that accounts for the self-interaction of the exogenous construct. The product indicator approach, the orthogonalizing approach, and the two-stage approach are three useful approaches to model the interaction term. The product indicator and orthogonalizing approaches are restricted to setups where the exogenous latent variable is measured reflectively. The two-stage approach can be used when formative or reflective measures are involved. The orthogonalizing approach proves valuable when the objective of the analysis is to minimize estimation bias of the quadratic effect or to maximize prediction. However, when the objective is to identify the statistical significance of the quadratic effect, the two-stage approach is preferred. Generally, the two-stage approach is the most versatile approach and is preferred.

REVIEW QUESTIONS

1. What is an HCM? Describe each of the four different types of HCMs introduced in this chapter.

2. Which criteria are relevant in the assessment of the different HCM types?

3. What are the consequences when LOCs have a substantially different number of indicators?

4. Explain the concept of self-interaction in the context of modeling quadratic effects.

5. What is the most versatile approach for creating a quadratic term with regard to the measurement model of the exogenous construct?

6. How do you assess the results of a nonlinear model?

CRITICAL THINKING QUESTIONS

1. Can every construct be measured at different levels of abstraction?

2. When would you use the two-stage HCM approach?

3. How do HCMs help solve discriminant validity problems in PLS-SEM?

4. Give an example of a cubic effect.

5. Explain what the path coefficients in a nonlinear model mean.

6. Discuss alternatives of analyzing a nonlinear effect in PLS-SEM.

KEY TERMS

Bandwidth-fidelity tradeoff

Bottom-up approach

Collect model

Cubic effect

Direct effect

Formative-formative HCM

Formative-reflective HCM

HCM

Hierarchical common factor model

Hierarchical component model (HCM)

Higher-order construct (HOC)

Higher-order model

HOC

Indirect effect

Interaction term

Jangle fallacy

Linear effect

Linear relationship

LOC

Log transformation

Lower-order construct (LOC)

Mediation

Mediator model

Moderation

Multiple battery model

Nonlinear effect

Nonlinear relationship

Orthogonalizing approach

Polynomial

Polynomial degree

Premature convergence

Product indicator approach

Product indicators

Quadratic effect

Reflective-formative HCM

Reflective-reflective HCM

Repeated indicators approach for HCM

Second-order construct

Self-interaction

Spread model

Stand-alone HCM

Top-down approach

Total effect

Total effects analysis of collect-type HCMs

Two-stage approach (nonlinear effects)

Two-stage approach for HCMs

SUGGESTED READINGS

Becker, J.-M., Klein, K., & Wetzels, M. (2012). Hierarchical latent variable models in PLS-SEM: Guidelines for using reflective-formative type models. *Long Range Planning, 45,* 359–394.

Henseler, J., Fassott, G., Dijkstra, T. K., & Wilson, B. (2012). Analysing quadratic effects of formative constructs by means of variance-based structural equation modelling. *European Journal of Information Systems, 21,* 99–112.

Rigdon, E. E., Ringle, C. M., & Sarstedt, M. (2010). Structural modeling of heterogeneous data with partial least squares. In N. K. Malhotra (Ed.), *Review of marketing research* (pp. 255–296). Armonk, NY: M. E. Sharpe.

Ringle, C. M., Sarstedt, M., & Straub, D. W. (2012). A critical look at the use of PLS-SEM in *MIS Quarterly. MIS Quarterly, 36,* iii–xiv.

Wetzels, M., Odekerken-Schroder, G., & van Oppen, C. (2009). Using PLS path modeling for assessing hierarchical construct models: Guidelines and empirical illustration. *MIS Quarterly, 33,* 177–195.

Wilson, B. (2010). Using PLS to investigate interaction effects between higher order branding constructs. In V. Esposito Vinzi, W. W. Chin, J. Henseler, & H. Wang (Eds.), *Handbook of partial least squares: Concepts, methods and applications* (pp. 621–652). New York, NY: Springer.

CHAPTER 3

Advanced Model Assessment

LEARNING OUTCOMES

1. Know how to assess the mode of a measurement model with the confirmatory tetrad analysis in PLS-SEM (CTA-PLS).

2. Understand how to run the CTA-PLS in SmartPLS and how to interpret the results.

3. Appreciate the importance-performance map analysis (IPMA).

4. Comprehend how to conduct the IPMA using the SmartPLS software.

CHAPTER PREVIEW

A fundamental concern in any partial least squares structural equation modeling (PLS-SEM) analysis is avoiding measurement model misspecification (e.g., specifying the measurement model as reflective when it should be formative). Any failure to specify the measurement models correctly can result in inaccurate estimates of the parameters. To address the issue of proper measurement model specification, this chapter introduces the confirmatory tetrad analysis for PLS-SEM (CTA-PLS). Drawing on an a priori measurement specification based on theory and logic (please also refer to the measurement model operationalization discussed in Chapter 1), the CTA-PLS provides a foundation for empirically assessing whether data support a formative measurement model specification or a reflective specification. Recent research calls for routine use of CTA-PLS to avoid measurement model misspecification in general (e.g., Bollen & Diamantoloulos, 2017) and PLS-SEM in particular (e.g., Hair, Sarstedt, Ringle, & Mena, 2012).

Another concern when applying PLS path modeling is how to best present the findings of your structural model. An excellent approach to interpret and explain your findings is importance-performance map analysis (IPMA). The second topic we cover in this chapter extends the standard results reporting of path coefficient estimates, R^2, and similar parameters, and adds a visual interpretation based on the average values of the latent variable scores. More precisely, IPMA contrasts the total effects, representing a predecessor constructs' influence on a specific target construct, with their average latent variable scores, thereby indicating the predictive strength of antecedent constructs. The analysis enables identification of predecessor (antecedent) constructs that have a relatively high (low) importance in predicting the target construct (i.e., those that have a strong [weak] total effect), but also have a relatively low (high) performance (i.e., low [high] average latent variable scores). These findings are particularly useful for interpreting where to emphasize efforts to improve the performance of a key target construct predicted in the PLS path model.

CONFIRMATORY TETRAD ANALYSIS

Method

Measurement model misspecification is a threat to the validity of SEM results (e.g., Jarvis, MacKenzie, & Podsakoff, 2003). For example, modeling latent variables reflectively (i.e., with a **reflective measurement model**) when the conceptualization of the measurement model, and thus the item wordings, follows a formative specification (i.e., a **formative measurement model**) can result in biased outcomes. The reason is that formative indicators are not necessarily correlated and are often not highly correlated. In addition, formative indicators produce lower outer loadings when represented in a reflective measurement model. Since indicators with lower outer loadings (<0.40) should be eliminated from reflectively measured latent constructs (see Hair, Hult, Ringle, & Sarstedt, 2017, Chapter 4), the incorrect specification of a measurement model as reflective when it should be formative can result in the deletion of indicators that should actually be retained. Any attempt to purify formative measurement models based on correlation patterns among the indicators can have adverse consequences for the constructs' content validity (e.g., Diamantopoulos & Siguaw, 2006) as such an (incorrect) indicator elimination typically has a significant impact on measurement and structural model results.

The primary means to decide whether to specify a measurement model reflectively or formatively is by theoretical reasoning. Guidelines such as Jarvis et al.'s (2003) decision rules (see also Hair et al., 2017, Chapter 2; and Chapter 1 in this book) prove helpful in this respect (Exhibit 3.1). Similarly, Bollen and Diamantopoulos (2017) provide helpful guidance:

> Mental experiments are one conceptual tool for determining the nature of an indicator. A researcher should imagine a change in the indicator and ask whether this change is likely to change the value of the latent variable. If so, this is theoretical evidence supporting causal or formative indicators. Alternatively, the researcher should imagine changing the latent variable and ask whether this is likely to change the value of the indicator(s). If so, this favors reflective (effect) indicators. Such mental experiments can provide conceptual support for one type of indicator over the other.

While substantiation of the theoretically and conceptually established measurement model is useful to avoid measurement model misspecification (Bollen & Diamantopoulos, 2016), there are also empirical means that can help researchers understanding whether to specify a measurement model reflectively or formatively. Specifically, Bollen and Ting (1993, 2000) introduced the confirmatory tetrad analysis, which Gudergan, Ringle, Wende, and Will (2008) adapted to PLS-SEM. The **confirmatory tetrad analysis in PLS-SEM (CTA-PLS)** enables researchers to empirically evaluate whether the measurement model specification chosen based on theoretical grounds is supported by the data (Rigdon, 2005). More precisely, CTA-PLS can confirm the appropriateness of a reflective measurement model specification. On the other hand, when the method disconfirms the appropriateness of a reflective measurement model, it provides support for a formative measurement model specification. That is, the method enables

Exhibit 3.1   Decision Rules for Determining Whether a Construct Is Formative or Reflective

	Formative Model	Reflective Model
1. Direction of causality from construct to measure implied by the conceptual definition	Direction of causality is from items to construct.	Direction of causality is from construct to items.
Are the indicators (items) (a) defining characteristics or (b) manifestations of the construct?	Indicators are defining characteristics of the construct.	Indicators are manifestations of the construct.
Would changes in the indicators/items cause changes in the construct?	Changes in the indicators should cause changes in the construct.	Changes in the indicator should not cause changes in the construct.
Would changes in the construct cause changes in the indicators?	Changes in the construct do not cause changes in the indicators.	Changes in the construct do cause changes in the indicators.
2. Interchangeability of the indicators/items	Indicators need not be interchangeable.	Indicators should be interchangeable.
Should the indicators have the same or similar content? Do the indicators share a common theme?	Indicators need not have the same or similar content; indicators need not share a common theme.	Indicators should have the same or similar content; indicators should share a common theme.
Would dropping one of the indicators alter the conceptual domain of the construct?	Dropping an indicator may alter the conceptual domain of the construct.	Dropping an indicator should not alter the conceptual domain of the construct.
3. Covariation among the indicators	It is not necessary for indicators to covary with each other.	Indicators are expected to covary with each other.
Should a change in one of the indicators be associated with changes in the other indicators?	Not necessarily.	Yes.
4. Nomological net of the construct indicators	Nomological net for the indicators may differ.	Nomological net for the indicators should not differ.
Are the indicators/items expected to have the same antecedents and consequences?	Indicators are not required to have the same antecedents and consequences.	Indicators are required to have the same antecedents and consequences.

Source: Jarvis et al. (2003).

substantiating the direction of the measurement model relationships or provides support for an alternative specification. Switching the measurement mode of latent variables (e.g., from a reflective one to a formative one, and vice versa) solely on the grounds of CTA-PLS results is, however, not meaningful unless additional theoretical or conceptual logic provides support for this change.

The CTA-PLS builds on the concept of **tetrads** (τ), which describe the relationship between pairs of covariances. To better understand what a tetrad is, consider a reflectively measured latent variable Y_1 with four indicators x_1 to x_4. For this construct, we obtain six covariances (σ) between all possible pairs of the four indicators, as shown in Exhibit 3.2.

A tetrad is the difference of the product of one pair of covariances and the product of another pair of covariances. The six covariances of four indicator variables result in six unique pairings that form three tetrads:

$$\tau_{1234} = \sigma_{12} \cdot \sigma_{34} - \sigma_{13} \cdot \sigma_{24},$$

$$\tau_{1342} = \sigma_{13} \cdot \sigma_{42} - \sigma_{14} \cdot \sigma_{32}, \text{ and}$$

$$\tau_{1423} = \sigma_{14} \cdot \sigma_{23} - \sigma_{12} \cdot \sigma_{43}.$$

In reflective measurement models, each tetrad is expected to have a value of zero and, thereby, to vanish—these **vanishing tetrads** are used as input for the analyses. The reason is that, according to the domain sampling model (e.g., Nunnally & Bernstein, 1994), reflective indicators represent one specific concept or trait equally well.

Exhibit 3.2 **Covariances of Four Indicators in a Reflective Measurement Model**

	x_1	x_2	x_3	x_4
x_1				
x_2	σ_{12}			
x_3	σ_{13}	σ_{23}		
x_4	σ_{14}	σ_{24}	σ_{34}	

Therefore, differences between pairs of covariances of indicators that represent the concept in a similar manner should be zero, provided the domain sampling model holds as assumed by a reflective measurement model. If only one tetrad value in a measurement model is significantly different from zero (i.e., it does not vanish), one has to reject the reflective measurement model specification and, instead, assume the alternative formative specification. In other words, the CTA-PLS is a statistical test that considers the hypothesis $H_0 : \tau = 0$ (i.e., the tetrad equals zero and vanishes) and the alternative hypothesis $H_1 : \tau \neq 0$ (i.e., the tetrad does not equal zero). That is, the CTA-PLS initially assumes a reflective measurement specification. However, a significant test statistic provides support for H_1, suggesting that the measurement model is formative.

Bollen and Ting (2000) provide several numerical examples that illustrate the usefulness of the CTA in the context of covariance-based SEM (CB-SEM). Although the procedures differ, the systematic application of the CTA for assessing measurement models in PLS-SEM is similar to its CB-SEM counterpart (Bollen & Ting, 2000). A necessary requirement for any CTA-PLS is that the items are correlated—at least to some degree. If they were uncorrelated, all tetrads would by definition be zero, which would render the CTA-PLS meaningless. Therefore, a CTA-PLS first requires testing whether at least some of the measurement model's indicators are significantly correlated, for example, by using the IBM SPSS Statistics software (for more details, see Sarstedt & Mooi, 2014). If this requirement is met, CTA-PLS involves the following main steps:

1. Form and compute all tetrads for the measurement model of a latent variable.

2. Identify and eliminate redundant tetrads.

3. Perform a statistical significance test whether each tetrad vanishes.

4. Evaluate whether a measurement model's nonredundant tetrads vanish.

In Step 1, all tetrads of the latent variables' measurement models are computed. A key consideration is that the tetrad construction requires at least four indicators per measurement model. Otherwise, the CTA-PLS will not produce results for the construct. Bollen and

Ting (2000) and Gudergan et al. (2008) provide some additional advice regarding how to deal with situations of less than four indicators (i.e., two and three per measurement model), but in PLS-SEM we strongly recommend applying the CTA only on measurement models with at least four indicators. In general, there will be $m!/\left((m-4)!4!\right)\cdot 3$ tetrads for measurement models with m indicators (Bollen & Ting, 2000). For example, a measurement model with five, eight, or ten indicators yields 15, 210, or 630 tetrads.

Step 2 of the CTA-PLS identifies and eliminates redundant tetrads. Redundancy exists whenever a tetrad can be represented by two other tetrads. For example, consider the three tetrads extracted from Exhibit 3.2 (i.e., $\tau_{1234}, \tau_{1342}, \tau_{1423}$). Each tetrad can be represented by the algebraic combination of the other two tetrads (Bollen & Ting, 1993). Thus, one tetrad is redundant and can be eliminated (e.g., τ_{1423}), which leaves two **nonredundant tetrads** for the remaining analysis (e.g., τ_{1234} and τ_{1342}).

While the first two steps of CTA-PLS deal with generating and selecting the (nonredundant) tetrads for each measurement model, Steps 3 and 4 address their significance testing. In Step 3, the CTA-PLS draws on bootstrapping to test whether the nonredundant tetrads differ significantly from zero (see Hair et al., 2017, Chapter 5, for further information on the bootstrapping procedure and recommendations for algorithm settings). Analyzing a large number of nonredundant tetrads using bootstrapping requires running a great number of tests. Not only do such complex analyses become increasingly time consuming, but there also is a more severe problem associated with this approach, called **alpha inflation** (also referred to as **multiple testing problem**). This term refers to the fact that the more tests you conduct at a certain significance level, the more likely you are to claim a significant result when this is not so (i.e., a Type I error). The greater the number of nonredundant tetrads in a particular measurement model, the higher the likelihood that a rejection of the null hypothesis (i.e., a tetrad's value significantly differs from zero) will occur just by chance. For this reason, in Step 4 the CTA-PLS applies a Bonferroni correction that adjusts for an alpha inflation. As a result of the Bonferroni correction, the tetrad tests in each measurement model do not assume a significance level of alpha (typically 10%) but alpha divided by the number of nonredundant tetrads. For example, with 10 nonredundant tetrads in a measurement model, each test would assume a significance level of $0.05 / 10 = 0.005$. Step 4 calculates the

bias-corrected and Bonferroni-adjusted confidence intervals of the nonredundant tetrads for a prespecified error level.

A nonredundant tetrad is significantly different from zero if its bias-corrected and Bonferroni-adjusted confidence interval does not include zero. That is, if a measurement model has only vanishing tetrads, we cannot reject the null hypothesis and assume a reflective measurement model specification. On the other hand, if only one of a measurement model's nonredundant tetrads is significantly different from zero (i.e., not all tetrads are vanishing tetrads), one should consider a formative measurement model specification. In any case, it is important to note that if the CTA-PLS does not support a reflective measurement model, adjustments in the model must be consistent with theoretical/conceptual considerations and not solely draw on the empirical test results.

Exhibit 3.3 shows an example of two reflectively measured latent variables Y_1 and Y_2 with four and five indicators, respectively. The measurement model of Y_1 with four indicators requires analyzing two nonredundant tetrads, while the measurement model of Y_2 with five indicators requires considering five nonredundant tetrads.

Exhibit 3.4 shows the (vanishing) tetrads and gives an example of their values as well as the bootstrapping results for 5,000 samples (bootstrap standard error, t value, and p value of every single tetrad). The tetrad values τ_{1235} and τ_{1352} are significantly different from zero. However, these results do not account for the multiple testing problem. For this reason, the last column in Exhibit 3.4 shows the 90% bias-corrected and Bonferroni-adjusted bootstrap confidence

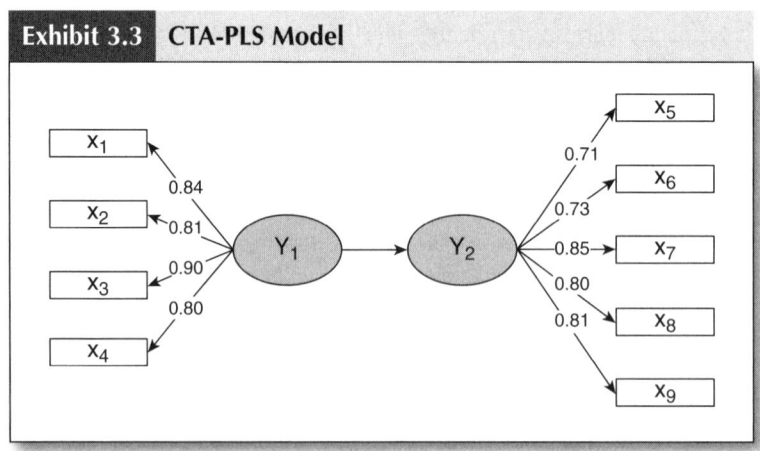

Exhibit 3.3 **CTA-PLS Model**

intervals. A tetrad may be significantly different from zero when analyzed in isolation but when accounting for the multiple testing problem it may be not significant based on the 90% bias-corrected and Bonferroni-adjusted bootstrap confidence intervals. In the example shown in Exhibit 3.4, we find that the value of two tetrads of Y_2 (i.e., τ_{1235} and τ_{1352}) are significant as zero does not fall into the 90% bias-corrected and Bonferroni-adjusted bootstrap confidence intervals (i.e., these two tetrads do not vanish). The value of tetrad τ_{1235} is significantly higher than zero, while the value of tetrad τ_{1352} is significantly lower than zero. Hence, the CTA-PLS results suggest that Y_2 should be specified as being formative. Note again that this result does not mean that one should mechanically switch to a formative specification, as any change in measurement must be substantiated by theoretical considerations. In contrast to Y_2, all tetrads of Y_1's measurement model are not significantly different from zero, which empirically substantiates the reflective measurement model specification.

Exhibit 3.4					Example CTA-PLS Results
Y_1	Tetrad value	Bootstrap standard error	t value	p value	$CI_{adj.}{}^a$
τ_{1234}	0.194	0.150	1.298	0.194	[−0.099; 0.488]
τ_{1243}	−0.115	0.182	0.632	0.527	[−0.469; 0.245]
Y_2	Tetrad value	Bootstrap standard error			$CI_{adj.}{}^a$
τ_{1234}	0.159	0.139	1.140	0.254	[−0.165; 0.483]
τ_{1243}	0.223	0.145	1.538	0.124	[−0.116; 0.558]
τ_{1235}	0.483	0.142	3.408	0.001	[0.151; 0.811]
τ_{1352}	−0.346	0.121	2.856	0.004	[−0.626; −0.062]
τ_{1345}	−0.089	0.136	0.656	0.512	[−0.404; 0.230]

[a] $CI_{adj} = 90\%$ bias-corrected and Bonferroni-adjusted bootstrap confidence intervals.

Finally, we would like to express two notes of caution: First and foremost, CTA-PLS is no silver bullet and its results do not discharge researchers from closely thinking about the specification of measurement models. Changing the direction of the measurement models solely on the grounds of CTA-PLS results is unreasonable. Instead, the CTA-PLS results should be interpreted as inputs either to confirm the underlying theoretical and conceptual reasoning or to critically but cautiously reassess the measurement model specification subject to theoretical and conceptual logics. Second, the CTA-PLS results do not allow making conclusions about the content validity of the measurement models. That is, the results of CTA-PLS do not indicate whether sufficient parts of a construct domain have been captured by the indicators. This aspect of an empirical analysis requires careful reasoning based on theory and logic, perhaps qualitative research, and potentially with the support of experts who are knowledgeable about the conceptual domain.

The primary rules of thumb on how to conduct CTA-PLS are shown in Exhibit 3.5. In the following case study, we show how to apply the analysis to the corporate reputation model example using the SmartPLS software.

Case Study Illustration— Confirmatory Tetrad Analysis

The CTA-PLS method is implemented in the SmartPLS 3 software, facilitating its straightforward use. Note that SmartPLS also allows running CTA-PLS on both reflectively and formatively measured latent variables. In both cases, the null hypothesis assumes a reflective measurement model when conducting the statistical test. The CTA-PLS results either support or challenge the selected reflective or formative measurement mode. However, for the reasons cited above, the CTA-PLS in SmartPLS includes only latent variables that have four or more indicators.

Our illustration of the CTA-PLS draws on the corporate reputation model, which has three latent variables with at least four indicators: *CSOR*, *PERF*, and *QUAL*. To start the analysis, we need to inspect whether the indicators per measurement model have correlations very close to zero. In that case, the tetrad would be per definition zero and the analysis becomes meaningless. To obtain the indicator correlations, navigate to the **Corporate Reputation - Advanced PLS-SEM Book** project in

Exhibit 3.5	Rules of Thumb for Conducting the Confirmatory Tetrad Analysis

- Establish the mode of measurement (i.e., reflective or formative) on theoretical and conceptual grounds. To do so, also consider qualitative decision rules, such as the four by Jarvis et al. (2003), to support the choice of mode based on theoretical reasoning. Use the CTA-PLS to empirically substantiate your theoretical considerations.

- Consider constructs with at least four indicators in the CTA-PLS, if possible. Otherwise and only if absolutely necessary, draw on Bollen and Ting (2000) and Gudergan et al. (2008), who provide some additional advice regarding how to deal with situations of less than four indicators (i.e., two and three per measurement model).

- Inspect the correlations of indicators per measurement model. These correlations must be statistically significant. Otherwise, the tetrad is approximately zero per se and the CTA-PLS becomes meaningless.

- Generate all tetrads per measurement model and exclude the redundant ones. Analyze whether the nonredundant tetrads are significantly different from zero (i.e., whether they vanish).

- Since significance testing accounts for several nonredundant tetrads per measurement at the same time, use the Bonferroni correction to account for the multiple testing problem.

- For significance testing, run the CTA-PLS for a high number of bootstrapping subsamples (e.g., 5,000) and generate the 90% bias-corrected and Bonferroni-adjusted bootstrap confidence interval for each nonredundant tetrad.

- If the confidence interval of a measurement model's tetrad does not include zero, reject the reflective measurement model and assume a formative measurement model. Otherwise, if all of the confidence intervals include zero, the CTA-PLS results empirically provide support for reflective measurement model specification.

- CTA-PLS results offer guidance regarding the correct measurement model specification, but the final decision of the mode of measurement model always builds on theoretical and conceptual considerations.

the **Project Explorer** and double-click on **Extended model.** Next, navigate to **Calculate** → **PLS Algorithm,** which you can find at the top of the SmartPLS screen. Alternatively, you can left-click on the wheel symbol with the label **Calculate** in the tool bar. A combo box opens from which you can select the **PLS Algorithm.** Retain the default settings (i.e., the path weighting scheme) as described in Hair et al. (2017) and click on **Start**

Calculation. Upon completion of the calculations, the results report automatically opens. When you click on **Indicator Data (Correlations)** under **Base Data** at the lower right side of the report, SmartPLS will display all indicator correlations. You can now inspect whether the indicator correlations of the *CSOR*, *PERF*, and *QUAL* constructs are sufficiently different from zero. To gain a better overview, you can also copy and paste the correlations into spreadsheet software such as Microsoft Excel by clicking on **Copy to Clipboard** (i.e., **Excel Format**) above the results table. The results show that the indicators have minimum correlations of 0.406 for *CSOR*, 0.261 for *PERF*, and 0.321 for *QUAL* (Exhibit 3.6). The values are clearly different from zero, so we can continue conducting the CTA-PLS. If in doubt, you can also run a separate correlation analysis using

Exhibit 3.6	**Indicator Correlations**							
	csor_1	csor_2	csor_3	csor_4	csor_5			
csor_1	1							
csor_2	0.406	1						
csor_3	0.491	0.455	1					
csor_4	0.415	0.473	0.502	1				
csor_5	0.522	0.448	0.540	0.452	1			
	perf_1	perf_2	perf_3	perf_4	perf_5			
perf_1	1							
perf_2	0.496	1						
perf_3	0.369	0.352	1					
perf_4	0.395	0.402	0.261	1				
perf_5	0.424	0.383	0.275	0.352	1			
	qual_1	qual_2	qual_3	qual_4	qual_5	qual_6	qual_7	qual_8
qual_1	1							
qual_2	0.496	1						
qual_3	0.509	0.530	1					
qual_4	0.461	0.507	0.591	1				
qual_5	0.532	0.399	0.632	0.532	1			
qual_6	0.493	0.425	0.601	0.568	0.604	1		
qual_7	0.424	0.353	0.474	0.502	0.545	0.483	1	
qual_8	0.462	0.321	0.384	0.347	0.387	0.381	0.342	1

statistical software programs such as IBM SPSS Statistics, Stata, or Statistica. We can now proceed with the CTA-PLS.

To run the CTA-PLS, click on **Calculate → Confirmatory Tetrad Analysis (CTA)**. In the dialogue box that opens (Exhibit 3.7), specify **5,000 subsamples** for the bootstrapping routine and select two-tailed testing (**Two Tailed**). We generally recommend choosing a **Significance Level** of **0.10** in a CTA-PLS. Click on the **Start Calculation** button in the lower right of the CTA-PLS dialogue box to run the analysis.

After the computation has finished, the results report automatically opens. Under **Final Results**, you can click on *CSOR*, *PERF*, and *QUAL*. For each latent variable, a results table appears that is similar to the one shown in Exhibit 3.8. Note that your results will slightly differ from those in Exhibit 3.8 since CTA-PLS builds on the bootstrapping routine that randomly draws the subsamples used for the analysis.

The **Original Sample (O)** column shows the results of the nonredundant tetrads. For example, for the latent variable *CSOR*, whose measurement model has five indicators, the analysis includes five nonredundant tetrads τ_{1234}, τ_{1243}, τ_{1235}, τ_{1352}, τ_{1345}. The number of tetrads considered for each latent variable in the analysis exponentially increases with the number of indicators per measurement model. While a measurement model with the minimum number of four indicators includes only two nonredundant tetrads, there are five tetrads for five indicators (*CSOR* and *PERF*) and 20 tetrads for *QUAL's* measurement model with eight indicators.

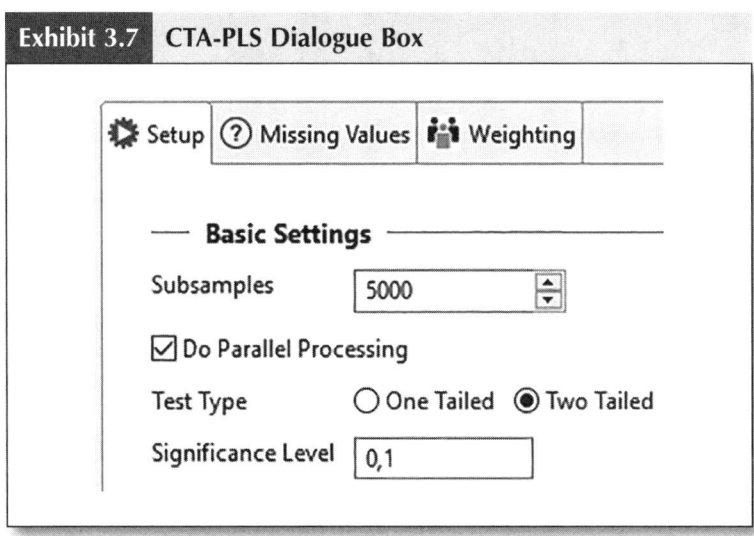

Exhibit 3.7   **CTA-PLS Dialogue Box**

Exhibit 3.8 CTA-PLS Results

CSOR	Original Sample (O)	Sample Mean (M)	Standard Deviation (STDEV)	T Statistics (|O/STDEV|)	p Values	Bias	CI Low	CI Up	Alpha adj.	z(1–alpha)	CI Low adj.	CI Up adj.
1: csor_1,csor_2,csor_3,csor_4	−0.148	−0.144	0.171	0.868	0.385	0.004	−0.425	0.137	0.02	2.327	−0.541	0.253
2: csor_1,csor_2,csor_4,csor_3	0.078	0.077	0.135	0.581	0.561	−0.002	−0.145	0.298	0.02	2.327	−0.237	0.390
4: csor_1,csor_2,csor_3,csor_5	−0.002	0.000	0.159	0.015	0.988	0.002	−0.261	0.261	0.02	2.327	−0.369	0.369
6: csor_1,csor_3,csor_5,csor_2	−0.101	−0.105	0.202	0.499	0.618	−0.005	−0.437	0.226	0.02	2.327	−0.575	0.364
10: csor_1,csor_3,csor_4,csor_5	−0.012	−0.014	0.142	0.087	0.931	−0.002	−0.248	0.219	0.02	2.327	−0.345	0.316

PERF	Original Sample (O)	Sample Mean (M)	Standard Deviation (STDEV)	T Statistics (|O/STDEV|)	p Values	Bias	CI Low	CI Up	Alpha adj.	z(1–alpha)	CI Low adj.	CI Up adj.
1: perf_1,perf_2,perf_3,perf_4	−0.073	−0.072	0.107	0.686	0.493	0.002	−0.248	0.104	0.02	2.327	−0.321	0.177
2: perf_1,perf_2,perf_4,perf_3	−0.038	−0.036	0.093	0.406	0.685	0.001	−0.189	0.116	0.02	2.327	−0.253	0.180
4: perf_1,perf_2,perf_3,perf_5	−0.019	−0.021	0.111	0.167	0.867	−0.002	−0.204	0.162	0.02	2.327	−0.280	0.238
6: perf_1,perf_3,perf_5,perf_2	−0.032	−0.031	0.097	0.333	0.739	0.002	−0.191	0.129	0.02	2.327	−0.257	0.196

PERF	Original Sample (O)	Sample Mean (M)	Standard Deviation (STDEV)	T Statistics (|O/STDEV|)	p Values	Bias	CI Low	CI Up	Alpha adj.	z(1-alpha)	CI Low adj.	CI Up adj.
10: perf_1,perf_3,perf_4,perf_5	0.091	0.091	0.126	0.720	0.472	0.000	-0.116	0.298	0.02	2.327	-0.202	0.384

QUAL	Original Sample (O)	Sample Mean (M)	Standard Deviation (STDEV)	T Statistics (|O/STDEV|)	p Values	Bias	CI Low	CI Up	Alpha adj.	z(1-alpha)	CI Low adj.	CI Up adj.
1: qual_1,qual_2,qual_3,qual_4	0.159	0.158	0.137	1.163	0.245	-0.001	-0.057	0.383	0.005	2.808	-0.226	0.542
2: qual_1,qual_2,qual_4,qual_3	0.223	0.220	0.139	1.596	0.111	-0.002	-0.009	0.450	0.005	2.808	-0.171	0.612
4: qual_1,qual_2,qual_3,qual_5	0.483	0.479	0.143	3.390	0.001	-0.004	0.245	0.714	0.005	2.808	0.079	0.880
6: qual_1,qual_3,qual_5,qual_2	-0.346	-0.346	0.121	2.846	0.004	0.000	-0.546	-0.146	0.005	2.808	-0.687	-0.004
7: qual_1,qual_2,qual_3,qual_6	0.384	0.380	0.143	2.683	0.007	-0.003	0.145	0.615	0.005	2.808	-0.021	0.782
10: qual_1,qual_2,qual_3,qual_7	0.266	0.264	0.137	1.937	0.053	-0.002	0.038	0.49	0.005	2.808	-0.121	0.649
13: qual_1,qual_2,qual_3,qual_8	0.116	0.117	0.111	1.049	0.294	0.000	-0.066	0.299	0.005	2.808	-0.194	0.428
17: qual_1,qual_2,qual_5,qual_4	-0.027	-0.030	0.179	0.151	0.880	-0.003	-0.324	0.264	0.005	2.808	-0.532	0.472
23: qual_1,qual_2,qual_7,qual_4	0.163	0.161	0.171	0.954	0.340	-0.003	-0.121	0.442	0.005	2.808	-0.321	0.642

(Continued)

Exhibit 3.8 (Continued)

QUAL	Original Sample (O)	Sample Mean (M)	Standard Deviation (STDEV)	T Statistics (|O/STDEV|)	p Values	Bias	CI Low	CI Up	Alpha adj.	z(1-alpha)	CI Low adj.	CI Up adj.
26: qual_1,qual_2,qual_8,qual_4	-0.274	-0.273	0.137	2.001	0.045	0.000	-0.498	-0.048	0.005	2.808	-0.657	0.111
30: qual_1,qual_5,qual_6,qual_2	0.137	0.136	0.131	1.042	0.298	-0.001	-0.080	0.352	0.005	2.808	-0.233	0.505
33: qual_1,qual_5,qual_7,qual_2	0.087	0.086	0.125	0.693	0.488	0.000	-0.119	0.292	0.005	2.808	-0.265	0.437
42: qual_1,qual_6,qual_8,qual_2	-0.172	-0.173	0.135	1.273	0.203	0.000	-0.396	0.050	0.005	2.808	-0.553	0.208
73: qual_1,qual_3,qual_7,qual_8	0.050	0.048	0.138	0.360	0.719	-0.002	-0.179	0.275	0.005	2.808	-0.340	0.436
85: qual_1,qual_4,qual_6,qual_7	-0.123	-0.125	0.149	0.824	0.410	-0.003	-0.370	0.120	0.005	2.808	-0.544	0.293
97: qual_1,qual_5,qual_6,qual_8	0.054	0.057	0.127	0.420	0.674	0.003	-0.153	0.267	0.005	2.808	-0.301	0.415
100: qual_1,qual_5,qual_7,qual_8	0.077	0.077	0.127	0.611	0.541	0.000	-0.131	0.286	0.005	2.808	-0.279	0.433
110: qual_2,qual_3,qual_6,qual_4	0.248	0.247	0.147	1.693	0.091	-0.001	0.006	0.488	0.005	2.808	-0.165	0.659
121: qual_2,qual_3,qual_5,qual_7	0.484	0.479	0.159	3.038	0.002	-0.006	0.216	0.741	0.005	2.808	0.031	0.926
156: qual_2,qual_6,qual_7,qual_5	0.095	0.093	0.158	0.603	0.546	-0.002	-0.167	0.352	0.005	2.808	-0.350	0.536

The bias-corrected and Bonferroni-adjusted confidence intervals indicate whether the nonredundant tetrads are significantly different from zero. **CI Low adj.** and **CI Up adj.** in the results report show the upper and the lower bounds of the 90% bias-corrected and Bonferroni-adjusted confidence intervals. If zero falls into the confidence interval, the tetrad is not significantly different from zero, which implies that it is a vanishing tetrad. Otherwise, if zero does not fall into the bias-corrected and Bonferroni-adjusted confidence interval, then the nonredundant tetrad is significantly different from zero. The latter situation occurs when the signs of **CI Low adj.** and **CI Up adj.** both are negative or both are positive. This is the case for tetrads 4, 6, and 121 of the *QUAL* construct. The numbers stem from the full list of tetrad generation (e.g., 210 tetrads for the measurement model of *QUAL* with eight indicators). This list becomes considerably reduced after determining the nonredundant tetrads in Step 2 of the CTA-PLS. SmartPLS, however, keeps the previous numbers to indicate which (nonredundant) tetrads have been selected for the final analysis. The confidence interval of *QUAL's* tetrad 4 has a lower boundary of 0.079 and an upper boundary of 0.880 and therefore does not include zero. Similarly, tetrad 121 has a significantly positive value. With a lower boundary of −0.687 and an upper boundary of −0.004, the confidence interval of tetrad 6 also does not include zero (i.e., it has a significant negative value). These results suggest that *QUAL's* measurement model is indeed formative, providing support for its original specification.

On the contrary, all nonredundant tetrads of *CSOR* and *PERF* vanish since all the tetrads' confidence intervals include zero. Hence, we assume that these constructs' measurement models are reflective, which opposes the theoretical and conceptual assumptions. In light of these results, we next examine the underpinnings of the measurement model specifications applying Jarvis et al.'s (2003) qualitative decision rules (Exhibit 3.1). Our examination of the construct indicators (Exhibit 3.9) suggests that these are rather independent sources that form the construct rather than reflections of some state. In other words, a change in the construct does not necessarily imply a simultaneous change of all indicators. Therefore, we retain the original measurement specification, also in light of the fact that various studies have used this specification in prior research (e.g., Eberl, 2010; Schwaiger, Sarstedt, & Taylor, 2010). Remember that the CTA-PLS results should not be mechanically applied, especially since formative indicators can exhibit strong correlations, which should, however, not be large (see Chapter 1).

Exhibit 3.9	Indicators of the Constructs Included in the CTA-PLS
Corporate Social Responsibility (CSOR)	
csor_1	[The company] behaves in a socially conscious way.
csor_2	[The company] is forthright in giving information to the public.
csor_3	[The company] has a fair attitude toward competitors.
csor_4	[The company] is concerned about the preservation of the environment.
csor_5	[The company] is not only concerned about profits.
Performance (PERF)	
perf_1	[The company] is a very well-managed company.
perf_2	[The company] is an economically stable company.
perf_3	The business risk for [the company] is modest compared to its competitors.
perf_4	[The company] has growth potential.
perf_5	[The company] has a clear vision about the future of the company.
Quality (QUAL)	
qual_1	The products/services offered by [the company] are of high quality.
qual_2	[The company] is an innovator, rather than an imitator with respect to [industry].
qual_3	[The company]'s products/services offer good value for money.
qual_4	The services [the company] offered are good.
qual_5	Customer concerns are held in high regard at [the company].
qual_6	[The company] is a reliable partner for customers.
qual_7	[The company] is a trustworthy company.
qual_8	I have a lot of respect for [the company].

To summarize, the CTA-PLS results indicate that $QUAL$'s measurement model should be specified as being formative whereas the measurement models of $CSOR$ and $PERF$ should be specified as being reflective. While these empirical findings match the theoretical and conceptual considerations with regard to $QUAL$, they oppose the original specification for the $CSOR$ and $PERF$ constructs. However, as the decision on how to specify measurement models primarily relies on theoretical and conceptual considerations, the CTA-PLS results should be taken as cause to critically reexamine, but not necessarily determine, these constructs' measurement model specifications. A reexamination of the conceptual foundations for these two constructs—employing the qualitative decisions rules of Jarvis et al. (2003)—however, supports the initial formative measurement model specification for $CSOR$ and $PERF$.

IMPORTANCE-PERFORMANCE MAP ANALYSIS

Overview

The **importance-performance map analysis** (**IPMA;** also called **importance-performance matrix**, **impact-performance map**, or **priority map analysis**) extends the standard PLS-SEM results reporting of path coefficient estimates and other parameters by adding a procedure that considers the average values of the latent variable scores (e.g., Fornell, Johnson, Anderson, Cha, & Bryant, 1996; Martilla & James, 1977; Ringle & Sarstedt, 2016; Slack, 1994). More precisely, the IPMA contrasts the total effects, representing the predecessor constructs' **importance** in predicting a specific target construct, with their average latent variable scores indicating their **performance**. The rational is to identify predecessor constructs that have a relatively high importance for predicting the target construct (i.e., those that have a strong **total effect**), but also have a relatively low performance (i.e., low average latent variable scores) so that improvements can be implemented. The following descriptions draw on Ringle and Sarstedt's (2016) tutorial on the method.

To illustrate the concept of an IPMA, consider the path model in Exhibit 3.10 with four constructs Y_1 to Y_4. In this path model, Y_4 represents the final target construct, directly predicted by Y_1, Y_2, and Y_3. Furthermore, Y_1 and Y_2 have **indirect effects** on Y_4 via Y_3. Adding the predecessor constructs' direct and indirect effects yields their total effects on Y_4, which represent the importance dimension in the IPMA. In contrast, these constructs' average latent variable scores represent their performance, in the sense that high values indicate a greater performance.

The IPMA combines these two aspects graphically by relating the (unstandardized) total effects on the x-axis with the latent variable scores, rescaled on a range from 0 to 100, on the y-axis. The result is a chart such as in Exhibit 3.11. For interpretation, we focus on constructs in the lower right area of the importance-performance map. These constructs have a high importance for the target construct such that they can have a strong impact but show a low performance. Consequently, there is a particularly high potential to improve the performance of the constructs positioned in this area. Conversely, the performance improvements priority of constructs with lower importance relative to the others is less pronounced. In Exhibit 3.10, Y_1 is particularly important for explaining the target construct Y_4. More precisely, ceteris paribus (i.e., when everything else remains constant), a one-unit point increase in Y_1's performance increases the performance of Y_4 by the value of Y_1's total effect on Y_4, which is 0.84. Since

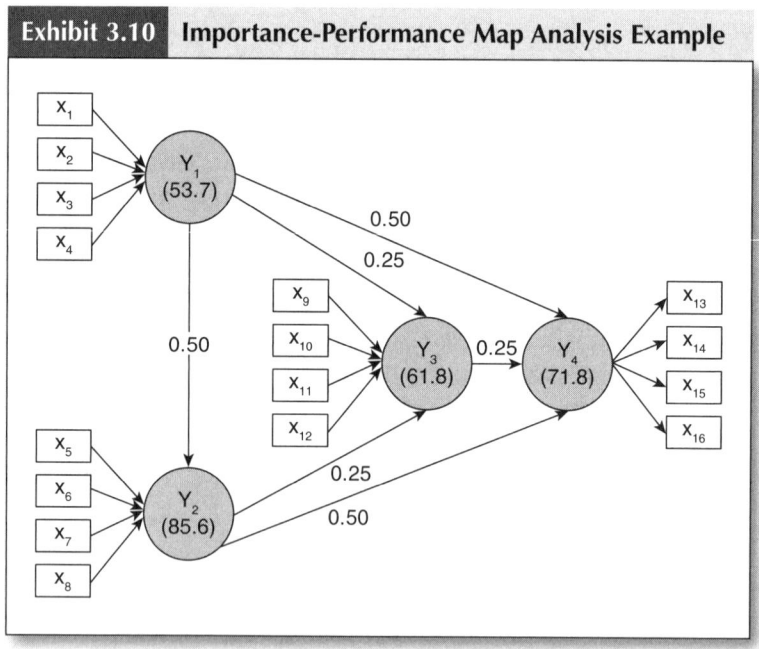

Exhibit 3.10   **Importance-Performance Map Analysis Example**

Source: Adapted from Ringle & Sarstedt (2016).

the performance of Y_1 is relatively low, there is substantial room for improvement, making the aspect underlying this construct particularly relevant for managerial or policy actions. While this introductory example shows an IPMA on the construct level, the analysis can also be run on the indicator level. In this case, individual data points in the importance-performance map are derived from indicator mean values and their total effect on a particular target construct.

Initial PLS-SEM studies show that IPMA results offer important insights into predecessor constructs' role and their relevance for managerial actions (e.g., Höck, Ringle, & Sarstedt, 2010; Kristensen, Martensen, & Grønholdt, 2000; Martensen & Grønholdt, 2010). The IPMA also is particularly useful when contrasting PLS-SEM results from a multigroup analysis (Chapter 4), as several studies illustrate (Rigdon, Ringle, Sarstedt, & Gudergan, 2011; Schloderer, Sarstedt, & Ringle, 2014; Völckner, Sattler, Hennig-Thurau, & Ringle, 2010).

The IPMA draws on the five-step procedure shown in Exhibit 3.12 (Ringle & Sarstedt, 2016). The first step involves checking whether the requirements for carrying out the analysis have been fulfilled (Step 1). The analysis proceeds with the computation of the latent variables' performance values (Step 2) and their importance values (Step 3). The importance-performance map creation for a selected

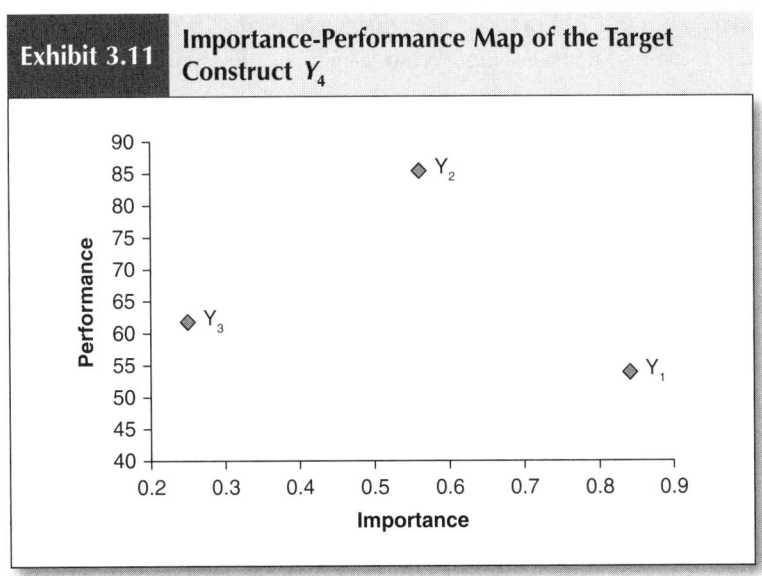

Exhibit 3.11	Importance-Performance Map of the Target Construct Y_4

Source: Adapted from Ringle & Sarstedt (2016).

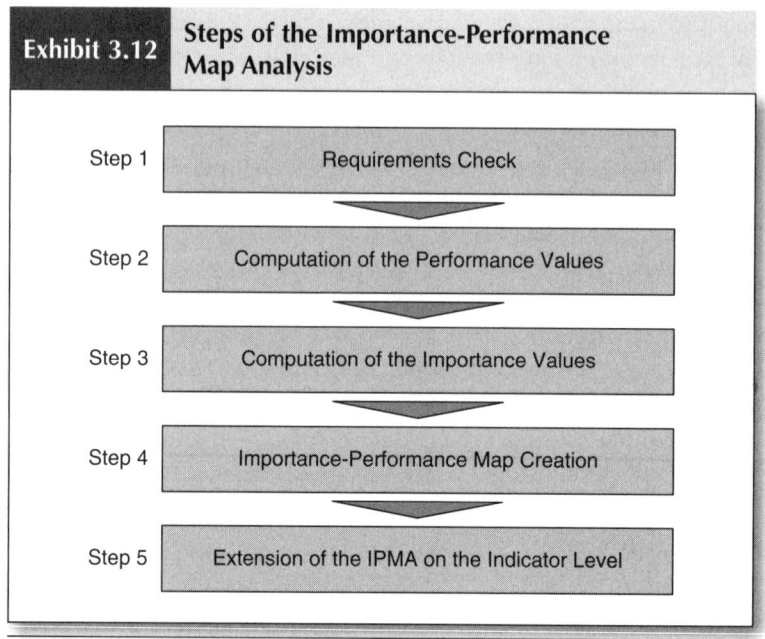

Exhibit 3.12	Steps of the Importance-Performance Map Analysis

Step 1 — Requirements Check

Step 2 — Computation of the Performance Values

Step 3 — Computation of the Importance Values

Step 4 — Importance-Performance Map Creation

Step 5 — Extension of the IPMA on the Indicator Level

Source: Adapted from Ringle & Sarstedt (2016).

target construct is based on these results (Step 4). Finally, the IPMA can be extended on the indicator level to obtain more specific information on those managerial or policy actions that can be most effective (Step 5). The following sections explain each step in greater detail.

Systematic IPMA Execution

Step 1: Requirements Check

IPMA applications have to meet three requirements. First, the **rescaling** of the latent variable scores on a range from 0 to 100 requires all indicators in the PLS path model to have a **metric scale** or at least an **equidistant scale**. In an equidistant scale, the intervals are distributed in equal units, which is typically the case when using **ordinal scales**. However, in order for an ordinal scale to be equidistant, the scale needs to be balanced, which means that there is an equal number of positive and negative categories. Therefore, **forced-choice scales** (i.e., scales without a neutral category) are, strictly speaking, not equidistant. For example, a 5-point Likert scale with two negative categories (completely disagree and disagree), a neutral option, and two positive categories (agree and completely agree) can be considered an equidistant scale. Indicators measured on a **nominal scale**, however, cannot be used in an IPMA. In a nominal scale, the assignments of numbers to certain object characteristics are interchangeable, rendering any rescaling arbitrary. Second, all the indicator coding must have the same scale direction. A low value on the scale must represent a negative or low outcome, and a high value must represent a positive or high outcome. If this requirement is not met, we cannot conclude that higher latent variable scores represent better performance. In this case, the indicator coding needs to be changed by reversing the scale (e.g., on a 5-point scale, 5 becomes 1 and 1 becomes 5, 4 becomes 2 and 2 becomes 4, and 3 remains unchanged). Third, regardless of a measurement model of the formative or reflective mode, the outer weights estimates must be positive. If the outer weights are negative, the latent variable scores will not fall within the 0 to 100 range, but would, for example, be between –5 and 95. Note that there are different reasons for (unexpected) negative outer weights. If an outer weight is negative and significant, the researcher should inspect the indicator and its scale. It may have another direction compared to the other indicators in the measurement model, which requires reversing the scale. In case of nonsignificant outer weights (with negative signs), the researcher may consider removing those indicators. Finally, negative outer weights might be a result of high indicator collinearity. For example, variance inflation factor (VIF) values of 5 and higher indicate a potential collinearity problem (Hair et al., 2017). In this case, the researcher may also consider removing indicators. However, removing indicators from measurement models involves some additional considerations as explained by Hair et al. (2017, Chapter 5) in greater detail.

While not being a formal requirement for running an IPMA, researchers should carefully consider PLS path model setups that favor IPMA use on the indicator level. When a latent variable of high priority for a specific target construct is identified, it is particularly advantageous to further analyze this predecessor construct's measurement model on the indicator level. Such an assessment is particularly useful when the measurement model is specified as formative (see Chapter 1)—as is the case in our sample model in Exhibit 3.10 (i.e., indicators x_1 to x_{12}). In this case, the indicators describe aspects that shape the corresponding construct, while their weights indicate each aspect's importance in this respect. Therefore, aspects underlying indicators with high weights should be given more attention to identify managerial or policy actions aimed at improving the target construct's performance (for an application, see Höck et al., 2010). Note that the IPMA can be applied on any kind of PLS path model, regardless of whether the latent variables' measurement models are formative or reflective. The IPMA builds on the outer weights—as explained in more detail in the subsequent sections—and PLS-SEM always provides outer weights estimates, also when a measurement model is specified as reflective.

Step 2: Computation of the Performance Values

The indicator data determine the latent variable scores and, thus, their performance. Similarly, when conducting an IPMA on the indicator level, the mean value of an indicator represents its average performance. When computing average values on the construct or indicator level, it is important to remember that indicators may be measured on different scales. For examples, some indicators may use a scale with values from 1 to 5, while others use a scale with values from 1 to 7 or from 1 to 9. To facilitate the interpretation and comparison of performance levels, the IPMA rescales indicator scores on a range between 0 and 100, with 0 representing the lowest and 100 representing the highest performance. Since most researchers are familiar with interpreting percentage values, this kind of performance scale is easy to understand. The rescaling of an observation j with respect to indicator i proceeds via

$$x_{ij}^{\text{rescaled}} = \frac{E[x_{ij}] - \min[x_i]}{\max[x_i] - \min[x_i]} \cdot 100,$$

where x_i is the i-th indicator in the PLS path model, $E[.]$ represents indicator i's actual score of respondent j, and $\min[.]$ and $\max[.]$ represent the indicator's minimum and maximum value. It is important to note that the minimum and maximum values refer to the potential values on a certain scale (e.g., 1 and 5 on a 1 to 5 scale) and not the minimum and maximum values of the actual responses (e.g., 2 and 4 on a 1 to 5 scale). For example, according to this formula, a value of 4 on a 1 to 5 scale becomes $(4-1)/(5-1) \cdot 100 = 75$. All data points used for estimating the PLS path model are rescaled this way.

Exhibit 3.13 shows an excerpt of the original indicator data $(n = 300)$ used to estimate the sample model from Exhibit 3.10. All indicators are measured on a scale from 1 to 5.

Exhibit 3.13	Original Indicator Data															
Case	x_1	x_2	x_3	x_4	x_5	x_6	x_7	x_8	x_9	x_{10}	x_{11}	x_{12}	x_{13}	x_{14}	x_{15}	x_{16}
1	5	2	1	5	5	2	4	3	3	3	4	5	1	2	4	3
2	4	5	2	3	4	1	1	2	5	4	5	3	3	5	4	3
3	4	1	3	4	5	3	5	3	5	3	4	5	3	5	3	1
4	1	3	2	1	1	4	1	3	3	2	2	4	5	1	4	4
5	3	2	1	3	5	2	2	3	5	1	2	3	1	3	4	5
...	...	...	...	...	...	...	...	...	...	...	...	...	...	...	...	...
299	2	4	3	4	4	2	2	1	4	3	3	4	2	5	4	3
300	2	4	3	1	1	1	4	3	5	2	3	5	4	2	1	5
Mean Value	4.2	4.1	3.4	2.3	3.6	4.5	3.4	3.1	3.4	4.7	4.4	4.6	3.4	4.2	4.5	4.2

Source: Adapted from Ringle & Sarstedt (2016).

Exhibit 3.14 Rescaled Indicator Data

Case	x_1	x_2	x_3	x_4	x_5	x_6	x_7	x_8	x_9	x_{10}	x_{11}	x_{12}	x_{13}	x_{14}	x_{15}	x_{16}
1	100	25	0	100	100	25	75	50	50	50	75	100	0	25	75	50
2	75	100	25	50	75	0	0	25	100	75	100	50	50	100	75	50
3	75	0	50	75	100	50	100	50	100	50	75	100	50	100	50	0
4	0	50	25	0	0	75	0	50	50	25	25	75	100	0	75	75
5	50	25	0	50	100	25	25	50	100	0	25	50	0	50	75	100
⋮	⋮	⋮	⋮	⋮	⋮	⋮	⋮	⋮	⋮	⋮	⋮	⋮	⋮	⋮	⋮	⋮
299	25	75	50	75	75	25	25	0	75	50	50	75	25	100	75	50
300	25	75	50	0	0	0	75	50	100	25	50	100	75	25	0	100
Mean Value	79.0	77.5	59.5	33.5	66.0	87.5	59.5	53.0	59.5	91.5	85.5	90.0	59.5	79.0	88.5	79.0

Source: Adapted from Ringle & Sarstedt (2016).

Exhibit 3.14 shows the indicator data from Exhibit 3.13, rescaled on a range from 0 to 100, which serve as input for the computation of the rescaled latent variable scores. In addition, the mean values of the rescaled indicators represent their performance values (e.g., 79 for indicator x_1 and 77.5 for indicator x_2), which are later used for the IPMA on the indicator level.

Irrespective of whether the measurement model of a latent variable is reflective (i.e., Y_4) or formative (i.e., Y_1, Y_2, and Y_3), the rescaled latent variable scores are a linear combination of the rescaled indicator data and the rescaled outer weights. Hence, the IPMA can be executed for reflective and formative constructs but uses only the outer weights in both cases. Even though the IPMA permits using both types of measurement models (as well as single item constructs), it is particularly useful in situations when the endogenous target construct has a reflective measurement model while the exogenous constructs have formative measurement models. In this situation, the performance improvements of indicators in the formative measurement models result in a performance increase of the exogenous constructs, which in turn translates into a performance enhancement of the key target construct (see Höck et al., 2010).

To obtain the rescaled weights, we must first compute the unstandardized weights by dividing each standardized weight by the standard deviation of its respective indicator. While the standardized outer weights originate from the standard PLS path model estimation, the estimation of each indicator's standard deviation is based on the original indicator data. For example, if x_1 has a standardized weight of 0.2 and a standard deviation of 1.619, the resulting unstandardized weight is 0.124. Exhibit 3.15 shows the (standardized and unstandardized) indicator weights along with the indicators' standard deviations with regard to our sample model in Exhibit 3.10.

Finally, we rescale the unstandardized outer weights so that their sum equals one per measurement model. For this purpose, we need to divide each indicator's unstandardized weight (e.g., 0.124 for indicator x_1) by the sum of the unstandardized weights of all the indicators that belong to the same measurement model. For Y_1, the sum of all the unstandardized indicator weights is $0.124 + 0.168 + 0.191 + 0.406 = 0.889$. Therefore, for indicator x_1, we obtain the unstandardized and rescaled outer weight of 0.139 after dividing 0.124 by 0.889. The final column in Exhibit 3.15 shows the results of the rescaled outer weights.

Exhibit 3.15 Computation of Unstandardized and Rescaled Outer Weights

Latent Variable	Indicator	Standardized Outer Weights	Standard Deviation of the Indicators	Unstandardized Outer Weights	Rescaled Outer Weights
Y_1	x_1	0.2	1.619	0.124	0.139
	x_2	0.3	1.789	0.168	0.189
	x_3	0.4	2.099	0.191	0.215
	x_4	0.5	1.231	0.406	0.457
Y_2	x_5	0.1	2.099	0.048	0.114
	x_6	0.1	1.101	0.091	0.218
	x_7	0.4	3.357	0.119	0.285
	x_8	0.4	2.504	0.160	0.383

Latent Variable	Indicator	Standardized Outer Weights	Standard Deviation of the Indicators	Unstandardized Outer Weights	Rescaled Outer Weights
Y_3	x_9	0.1	1.762	0.057	0.136
	x_{10}	0.1	1.744	0.057	0.137
	x_{11}	0.4	2.270	0.176	0.422
	x_{12}	0.4	2.653	0.151	0.361
Y_4	x_{13}	0.3	2.164	0.139	0.186
	x_{14}	0.3	1.874	0.160	0.215
	x_{15}	0.3	1.413	0.212	0.286
	x_{16}	0.3	1.291	0.232	0.313

Source: Adapted from Ringle & Sarstedt (2016).

In the next step, the IPMA uses the rescaled indicator data (Exhibit 3.14) and the rescaled outer weights (Exhibit 3.15) to compute the rescaled latent variable scores by means of simple linear combinations. For example, the first data point in the vector of Y_1's scores is

$$100 \cdot 0.139 + 25 \cdot 0.189 + 0 \cdot 0.215 + 100 \cdot 0.457 \approx 64.3.$$

Exhibit 3.16 shows the resulting latent variable scores along with their mean values. In our example, Y_1 has a mean value (i.e., performance) of 53.7, Y_2 of 85.6, Y_3 of 61.8, and Y_4 of 78.1. These results serve as input for the importance-performance map's performance dimension.

Step 3: Computation of the Importance Values

A construct's importance in terms of explaining another directly or indirectly linked (target) construct in the structural model is derived from the total effect of the relationship between these two constructs.

Exhibit 3.16	Computation of the Rescaled Latent Variable Scores			
	Rescaled Latent Variable Scores			
	Y_1	Y_2	Y_3	Y_4
1	64.3	76.3	63.3	42.5
2	57.6	75.4	18.2	67.9
3	55.5	82.0	76.5	45.1
4	14.8	47.0	26.9	63.5
5	34.5	37.7	43.3	63.5
. . .	. . .	. . .	. . .	. . .
299	62.7	62.4	22.9	63.3
300	28.4	69.4	47.1	50.6
Mean Value	53.7	85.6	61.8	78.1

Source: Adapted from Ringle & Sarstedt (2016).

The total effect is the sum of the direct and all the indirect effects in the structural model (Hair et al., 2017). For example, to determine the total effect of Y_1 on Y_4 (Exhibit 3.10), we have to consider the **direct effect** of the relationship between these two constructs (0.50) and the following three indirect effects via Y_2 and Y_3, respectively:

$$Y_1 \to Y_2 \to Y_4 = 0.50 \cdot 0.50 = 0.25$$
$$Y_1 \to Y_2 \to Y_3 \to Y_4 = 0.50 \cdot 0.25 \cdot 0.25 = 0.03125$$
$$Y_1 \to Y_3 \to Y_4 = 0.25 \cdot 0.25 = 0.0625$$

Adding up the individual indirect effects yields the total indirect effect of Y_1 on Y_4, which is approximately 0.34. Therefore, the total effect of Y_1 on Y_4 is 0.84 ($= 0.50 + 0.34$). This total effect expresses Y_1's importance in predicting the target construct Y_4. It is important to note that the IPMA draws on unstandardized effects to facilitate a ceteris paribus (i.e., when everything else remains constant) interpretation of the impact of predecessor constructs on the target construct. More precisely, by drawing on unstandardized effects, we can conclude that an increase in a certain predecessor construct's performance would increase the target construct's performance by the size of its unstandardized total effect.

Exhibit 3.17 summarizes all the total effects with respect to our target construct Y_4. Note that Y_3 does not have an indirect effect on Y_4. Therefore, its total effect equals the direct effect of 0.25. At this point, after computing the importance and performance values, all information required to draw the importance-performance map is available.

Exhibit 3.17	Direct, Indirect, and Total Effects in the IPMA		
Predecessor Construct	Direct Effect on Y_4	Indirect Effect on Y_4	Total Effect on Y_4
Y_1	0.50	0.34	0.84
Y_2	0.50	0.06	0.56
Y_3	0.25	—	0.25

Source: Adapted from Ringle & Sarstedt (2016).
Note: All effects denote unstandardized effects.

Step 4: Importance-Performance Map Creation

The IPMA focuses on one key target construct of interest in the PLS path model. Therefore, the first step in creating an importance-performance map requires selecting the target construct of interest. In our example in Exhibit 3.10, Y_4 represents such a key target construct. In order to create the importance-performance map of Y_4, we need to use the importance and performance values of Y_4's predecessor constructs (i.e., Y_1, Y_2, and Y_3). Exhibit 3.18 summarizes the values of this map's importance and performance dimensions, as obtained by the previous IPMA steps.

Scatterplotting the information shown in Exhibit 3.18 enables us to create an importance-performance map as shown in Exhibit 3.11. The x-axis represents the importance of Y_1, Y_2, and Y_3 for predicting the target construct Y_4, while the y-axis depicts the performance of Y_1, Y_2, and Y_3 in terms of their average rescaled latent variable scores. For a better orientation, researchers may also draw two additional lines in the importance-performance map: the mean importance value (i.e., a vertical line) and the mean performance value (i.e., a horizontal line) of the displayed constructs (Exhibit 3.19). With regard to our example, Y_1, Y_2, and Y_3 have a mean importance of 0.55 and a mean performance of 67.0 (Exhibit 3.18). These two additional lines divide the importance-performance map into four areas with importance and performance values below and above the average. Generally, when analyzing the importance-performance map, constructs in the lower right area (i.e., above-average importance and below-average performance) represent the greatest opportunity to achieve improvement, followed by the upper right, lower left, and, finally, the upper left areas.

Exhibit 3.18	Data of the Importance-Performance Map for Construct Y_4	
	Importance	*Performance*
Y_1	0.84	53.7
Y_2	0.56	85.6
Y_3	0.25	61.8
Mean Value	0.55	67.0

Source: Adapted from Ringle & Sarstedt (2016).

Exhibit 3.19	Adjusted Importance-Performance Map of Target Construct Y_4

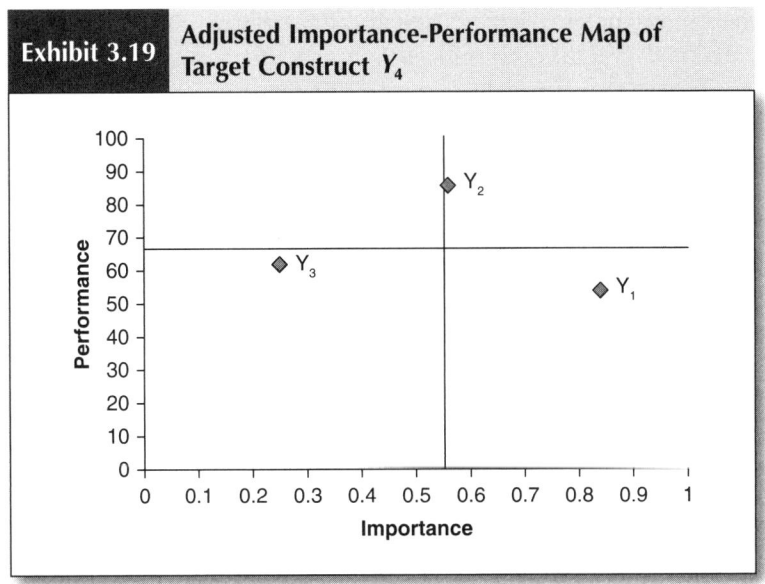

Source: Adapted from Ringle & Sarstedt (2016).

Thereby, the importance-performance map provides guidance for the prioritization of managerial or policy activities of high importance for the aspect underlying the selected target, but which require performance improvements.

In our example, the importance-performance map (Exhibit 3.19) shows that Y_1 has a relatively low performance of 53.7. In comparison with the other constructs, Y_1's performance is below average. On the other hand, with a total effect of 0.84, this construct's importance is particularly high. Therefore, a one-unit increase in Y_1's performance from 53.7 to 54.7 would increase the performance of Y_4 by 0.84 points from 78.10 to 78.94. Hence, when managers aim at increasing the performance of the target construct Y_4, their first priority should be to improve the performance of aspects captured by Y_1, as this construct has the highest (above average) importance, but a relatively low (below average) performance. Aspects related to constructs Y_2 and Y_3 follow as a second and third priority.

Step 5: Extension of the IPMA on the Indicator Level

The IPMA is not limited to the construct level. We can also conduct an IPMA on the indicator level to identify relevant and even more specific areas of improvement. More precisely, we can interpret the rescaled outer weights as an indicator's relative importance compared to that of the other indicators in a specific measurement model.

The importance values are derived from the indicators' total effects on the target construct, which is the result of multiplying the rescaled outer weights of a predecessor construct's indicators with its unstandardized total effect on the target construct. For example, with regard to the indicators of Y_1, we would multiply the rescaled outer weights of x_1 to x_4 (i.e., 0.139, 0.189, 0.215, 0.457; Exhibit 3.15) with the unstandardized total effect of Y_1 on Y_4 in the structural model (i.e., 0.84). This analysis yields importance values of x_1 to x_4 of, respectively, 0.117, 0.159, 0.181, and 0.384. The performance values are derived from the indicators' mean value of the rescaled data (i.e., 79, 77.5, 59.5, and 33.5; Exhibit 3.14). With this data for all indicators of Y_1, Y_2, and Y_3, we can create an importance-performance map as shown in Exhibit 3.20.

These results suggest that indicator x_4 should be given the highest priority for improvement, since it has the highest relative importance but the lowest performance. A one-unit point increase in x_4's performance increases the performance of Y_4 by x_4's importance value, which is 0.384 (ceteris paribus). Indicators x_8, x_3, x_7, and x_2 follow with second to fifth priority. The other indicators shown in

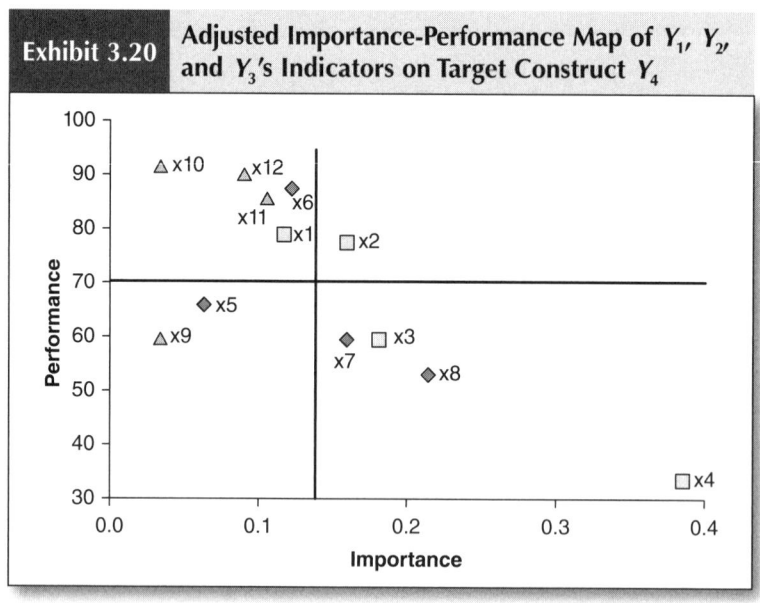

| Exhibit 3.20 | Adjusted Importance-Performance Map of Y_1, Y_2, and Y_3's Indicators on Target Construct Y_4 |

Source: Adapted from Ringle & Sarstedt (2016).

Exhibit 3.20 are less relevant for improving Y_4's performance. Instead of creating a single importance-performance map for all indicators (as shown in Exhibit 3.20), it is often more useful in applications to create separate importance-performance maps for the indicators of each construct (i.e., three indicator-level importance-performance map for constructs in our Y_1, Y_2, and Y_3 example).

Exhibit 3.21 summarizes the rules of thumb for conducting the IPMA.

Exhibit 3.21	Rules of Thumb for Carrying Out the Importance-Performance Map Analysis

- Only use indicators measured on a metric scale or equidistant scale (typically an ordinal or higher scale).

- All indicator coding must have the same scale direction (e.g., a low value on the scale must represent a negative or low outcome, and a high value must represent a positive or high outcome, as otherwise we cannot conclude that higher latent variable scores represent better performance).

- Regardless of the measurement model being specified as formative or reflective, the outer weights estimates must have the same (e.g., positive) sign to ensure interpretability of results.

- For the interpretability of outcomes, the total effects in the structural model should have the same (e.g., positive) signs.

- Select a key target construct and run the IPMA for its direct or all predecessor constructs in the structural model.

- Use the unstandardized total effects and the average latent variable scores on a scale from 0 to 100 to create the importance-performance map.

- Focus on those constructs with relatively low performance and a relatively high importance (i.e., a large total effect) on the target construct. The ceteris paribus interpretation of results assumes that an increase of a predecessor construct's performance by one point increases the performance of the selected key target construct by the size of the total effect.

Case Study Illustration— Importance-Performance Map Analysis

Our illustration of the IPMA draws on the corporate reputation model example. Exhibit 3.22 shows the model and the PLS-SEM results when using the empirical data and SmartPLS 3 software.

Exhibit 3.22 PLS-SEM and Results

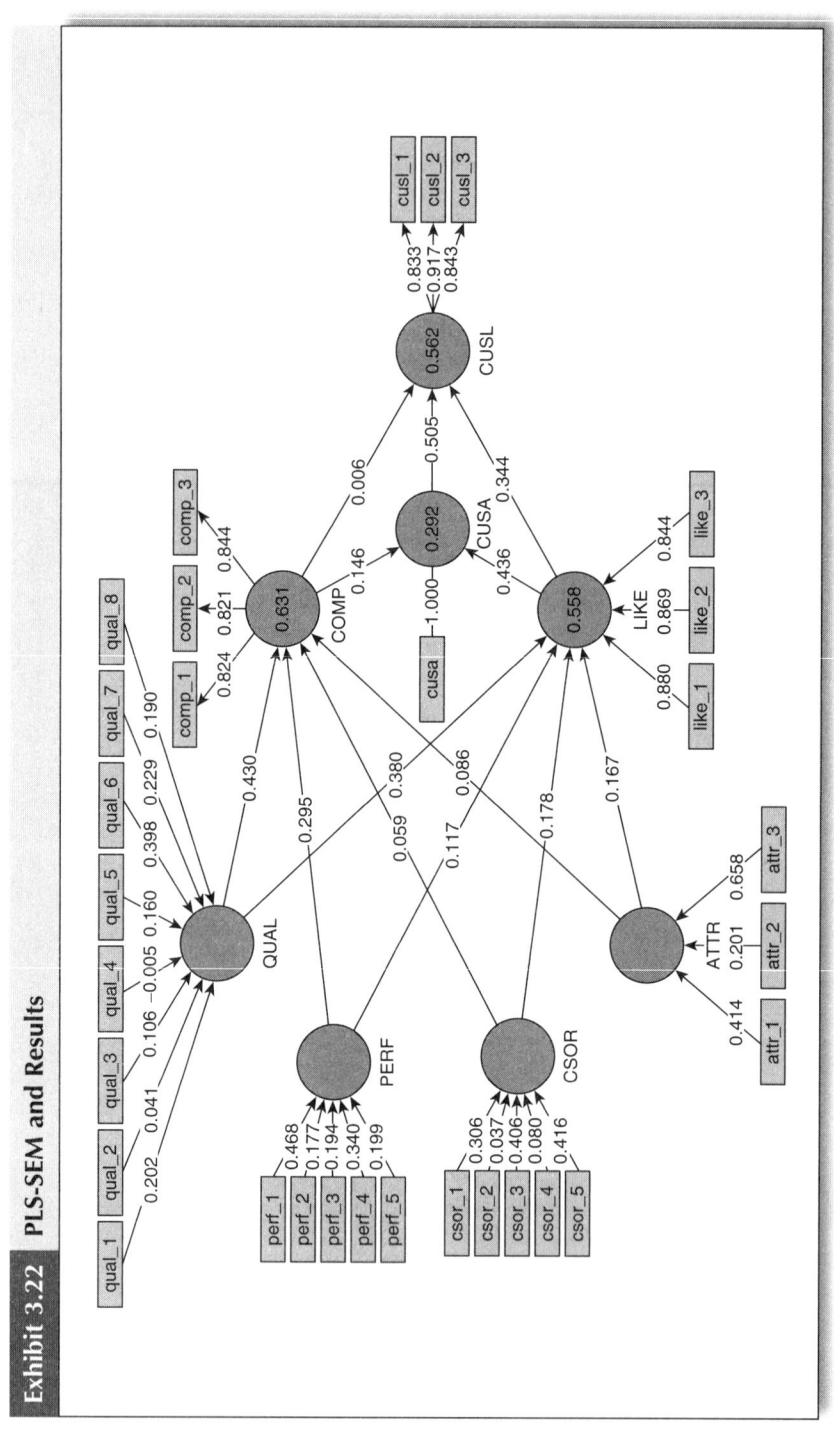

The results from bootstrapping with 5,000 samples using the no sign change option show that most of the path coefficients are statistically significant ($p < 0.01$). Only the path relationships $CSOR \rightarrow COMP$, $ATTR \rightarrow COMP$, and $COMP \rightarrow CUSL$ are not significant. We now extend the analysis and run the IPMA by following the procedure depicted in Exhibit 3.12. First, we check the requirements for carrying out an IPMA (Step 1). After reviewing the questionnaire, we find that the indicator data have been measured on a 7-point Likert scale with balanced categories and a neutral category (see Chapter 1). The indicators are measured on an equidistant scale. Furthermore, for all of the indicators, a higher value represents a better outcome (see Chapter 1). Therefore, we do not need to reverse-scale any of the indicators. To obtain further information on the data, double-click on **Corporate reputation data [344 records]** in the **Project Explorer** (Exhibit 3.23). The data view opens (Exhibit 3.24), which provides further information on the data set (e.g., the missing value marker) and some descriptive statistics.

Next, we inspect the signs of the outer weights. After running the PLS-SEM algorithm, view the **Results Report** (Exhibit 3.25) in SmartPLS, which also displays the weights of all the indicators. All the outer weights are positive, except for *qual_4*, whose weight is –0.005. However, this indicator weight is extremely close to zero and nonsignificant, so the negative weight's impact on the rescaling is extremely limited. Therefore, we retain this indicator and continue the analysis.

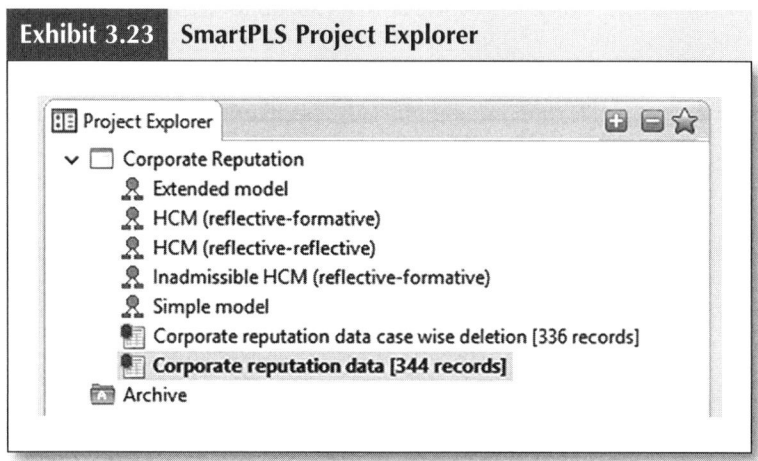

Exhibit 3.23 **SmartPLS Project Explorer**

Exhibit 3.24	SmartPLS Data View

Extended model.splsm | Corporate reputation data.txt

Delimiter:	Semicolon	Encoding:	UTF-8
Value Quote Character:	None	Sample size:	344
Number Format:	US (e.g. 1,000.23)	Indicators:	41
Missing Value Marker:	-99	Missing Values:	11

Indicators: | Indicator Correlations | Raw File

	No.	Missing	Mean	Median	Min	Max	Standard Deviation	Excess Kurtosis	Skewness
serviceprovider	1	0	2.000	2.000	1.000	4.000	1.003	-0.513	0.747
servicetype	2	0	1.637	2.000	1.000	2.000	0.481	-1.684	-0.571
comp_1	3	0	4.648	5.000	1.000	7.000	1.433	-0.324	-0.264
comp_2	4	0	5.424	6.000	1.000	7.000	1.375	-0.616	-0.566
comp_3	5	0	5.221	6.000	1.000	7.000	1.458	-0.188	-0.677
like_1	6	0	4.584	5.000	1.000	7.000	1.547	-0.399	-0.405
like_2	7	0	4.250	4.000	1.000	7.000	1.848	-0.901	-0.312
like_3	8	0	4.480	5.000	1.000	7.000	1.871	-0.941	-0.325
cusl_1	9	3	5.129	5.000	1.000	7.000	1.513	0.268	-0.792
cusl_2	10	4	5.276	6.000	1.000	7.000	1.744	0.040	-0.951
cusl_3	11	3	5.651	6.000	1.000	7.000	1.655	0.930	-1.301
cusa	12	1	5.440	6.000	1.000	7.000	1.174	0.777	-0.768
csor_1	13	0	4.235	4.000	1.000	7.000	1.469	-0.383	-0.042
csor_2	14	0	3.076	3.000	1.000	7.000	1.651	-0.564	0.497
csor_3	15	0	3.988	4.000	1.000	7.000	1.478	-0.468	-0.061
csor_4	16	0	3.125	3.000	1.000	7.000	1.462	-0.463	0.197

If the (negative) weight had been larger (e.g., –0.10), we would have had to delete this indicator for the IPMA. However, such a step has to be in line with measurement theory in that the remaining indicators still cover the construct domain sufficiently (see Hair et al., 2017, Chapter 5).

We subsequently run the IPMA by clicking on **Calculate** → **Importance-Performance Map Analysis (IPMA)** in the menu bar. Alternatively, you can left-click on the **Calculate** wheel symbol in the tool bar and select the corresponding option in the combo box that opens. In the following dialogue box (Exhibit 3.26), specify the target construct *CUSL* and choose the **All Predecessors of the Selected Target Construct** option. Most importantly, we need to specify each indicator's minimum and maximum value required for the rescaling of the data to a 0 to 100 scale. As shown in Exhibit 3.26, SmartPLS automatically reads these minimum and maximum values from the data. However, if the respondents have not made use of the full scale (e.g., the actual minimum value is 2 instead of 1), SmartPLS cannot rescale the data as you desire. If this is the case, change the minimum value manually by clicking on the corresponding cell in the **Min** or **Max** column and enter the correct minimum or maximum value of the scale. Alternatively, you can simultaneously specify the minimum

Exhibit 3.25	SmartPLS Results Report

Outer Weights

▦ Matrix

	ATTR	COMP	CSOR	CUSA	CUSL	LIKE	PERF	QUAL
attr_1	0.414							
attr_2	0.201							
attr_3	0.658							
comp_1		0.469						
comp_2		0.365						
comp_3		0.372						
csor_1			0.306					
csor_2			0.037					
csor_3			0.406					
csor_4			0.080					
csor_5			0.416					
cusa				1.000				
cusl_1					0.369			
cusl_2					0.420			
cusl_3					0.365			
like_1						0.419		
like_2						0.374		
like_3						0.363		
perf_1							0.468	
perf_2							0.177	
perf_3							0.194	
perf_4							0.340	
perf_5							0.199	
qual_1								0.202
qual_2								0.041
qual_3								0.106
qual_4								-0.005
qual_5								0.160
qual_6								0.398
qual_7								0.229
qual_8								0.190

and maximum value of all the indicators. To do so, enter the corresponding values next to **Min:/Max:** at the bottom of the dialogue box and click on **Apply to All**. In our example, all the respondents made use of the full range of the indicator scales as indicated in the **Min** and **Max** columns of the SmartPLS Data View (Exhibit 3.24). We therefore maintain the default settings (i.e., the path weighting scheme) and proceed by clicking on **Start Calculation**.

SmartPLS now automatically computes the performance and importance values (Steps 2 and 3) and creates the importance-performance map (Step 4). After completing the computations, SmartPLS opens the results report. Initially, it shows the results of the standardized path coefficients. Under **Quality Criteria → Importance-Performance Map [CUSL] (constructs, unstandardized effects)**, the report includes the importance-performance map as displayed in Exhibit 3.27. The graphical representation of the importance-performance uses the unstandardized total effects, which you can access by going to **Final Results → Total Effects** and clicking on the tab **Constructs, unstandardized**. Under **Final**

Exhibit 3.26 SmartPLS IPMA Dialogue

⚙ Setup ⚙ Partial Least Squares ⑦ Missing Values 👥 Weighting

— **Basic Settings** ————————————————————

Target Construct | CUSL ⌄ |

IPMA Results ◉ All Predecessors of the Selected Target Construct
○ Direct Predecessors of the Selected Target Construct

— **Ranges** ————————————————————

Manifest Variable	Latent Variable	Min	Max	^
attr_1	ATTR	1.0	7.0	
attr_2	ATTR	1.0	7.0	
attr_3	ATTR	1.0	7.0	
comp_1	COMP	1.0	7.0	
comp_2	COMP	1.0	7.0	⌄

Min: / Max: [   ] / [   ]   Apply to All   Reset

Exhibit 3.27	Importance-Performance Map (Construct Level)

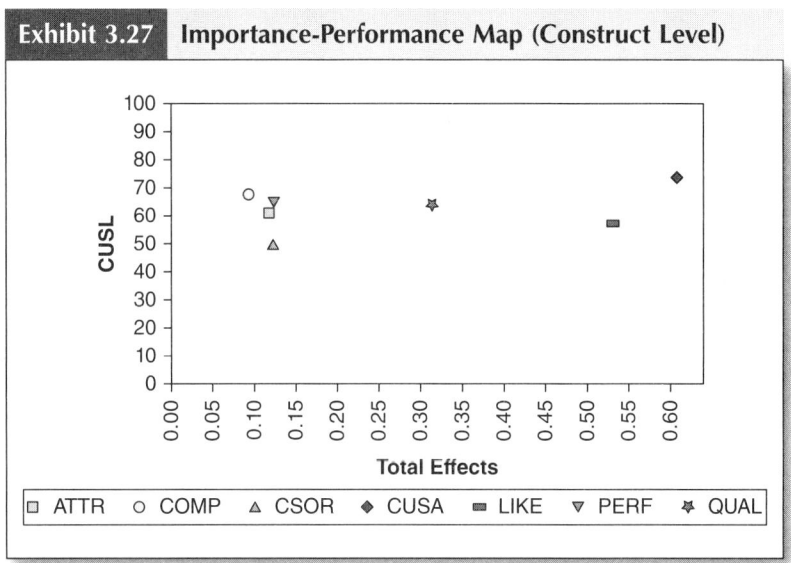

□ ATTR   ○ COMP   △ CSOR   ◆ CUSA   ▬ LIKE   ▽ PERF   ✿ QUAL

Results → Performance/Index, you can access the rescaled performance values of the latent and manifest variables (i.e., indicators) by clicking on the tabs **LV Performances** and **MV Performances.**

Not surprisingly, we find that two of the three direct predecessors, customer satisfaction (*CUSA*) and likeability (*LIKE*), have a particularly high importance for *CUSL* (Exhibit 3.27). More importantly, perceived quality (*QUAL*) has the greatest importance for *CUSL* compared to the other three exogenous constructs *ATTR*, *CSOR*, and *PERF*. Managerial or policy actions should therefore prioritize improving *CUSA* and *LIKE*, which can be achieved by focusing on the exogenous construct *QUAL*. Addressing *QUAL* is particularly promising as this construct has a formative measurement model, whose indicator weights indicate which aspect is most important for improving the respondents' perceived quality.

It is important to note that the graphical representation of the IPMA results differs from the graphical PLS-SEM results illustration in SmartPLS. Instead of displaying the R^2 values of the endogenous latent variables in the PLS path model (Exhibit 3.22), the IPMA results show the performance values of each latent variable (Exhibit 3.28); instead of displaying the standardized outer loadings or weights (Exhibit 3.22), the IPMA results show the unstandardized and rescaled outer weights of the measurement models regardless whether they are formative or reflective (Exhibit 3.28).

Exhibit 3.28 IPMA Results

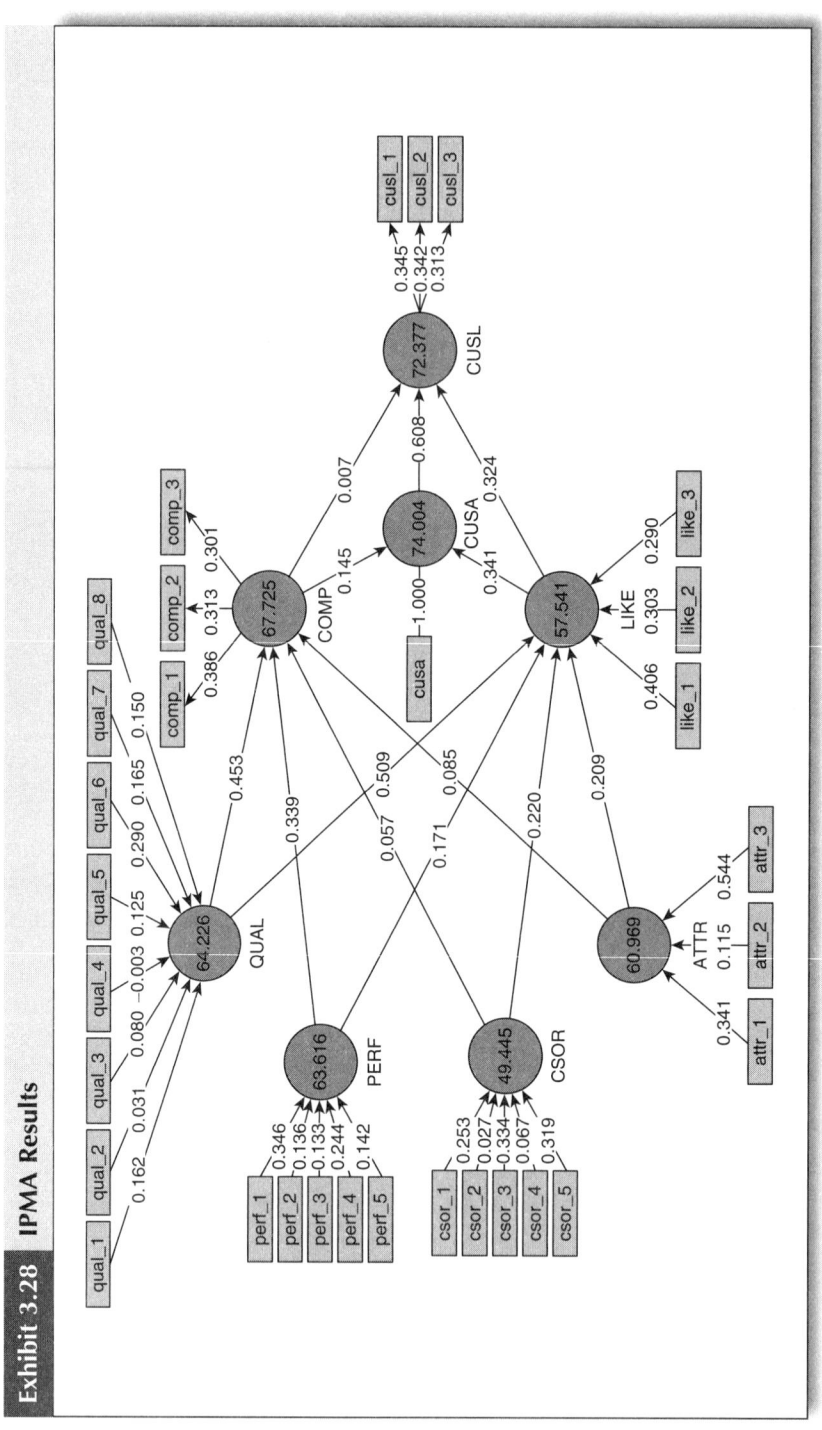

To gain more specific information on how to increase the performance of the constructs, particularly that of *QUAL*, the following analyses focus on the indicator level (Step 5). Note again that an IPMA on the indicator level is possible regardless of the predecessor constructs' measurement model specifications. However, an indicator-related analysis is particularly useful in formative measurement model settings. When going to **Quality Criteria → Importance-Performance Map [CUSL] (indicators, unstandardized effects)** in the results report, SmartPLS shows the indicators' importance-performance map as displayed in Exhibit 3.29. However, this display of all indicators does not offer helpful guidance for identifying features that should be addressed in order to improve customer loyalty. For this reason, we turn our attention toward the formatively measured constructs in the PLS path model as these represent relatively concrete features that can be addressed for performance improvements. Among these constructs, *QUAL* clearly has the greatest importance while showing an average performance. Hence, we focus on the IPMA of *QUAL*'s indicators.

Under **Quality Criteria → Importance-Performance Map [CUSL] (indicators, unstandardized effects)** in the results report, SmartPLS shows the indicators' unstandardized total effects in the **Indicator Total Effects for [CUSL]** tab and the performance values in the **Indicator Performances for [CUSL]** tab. Exhibit 3.30 shows these

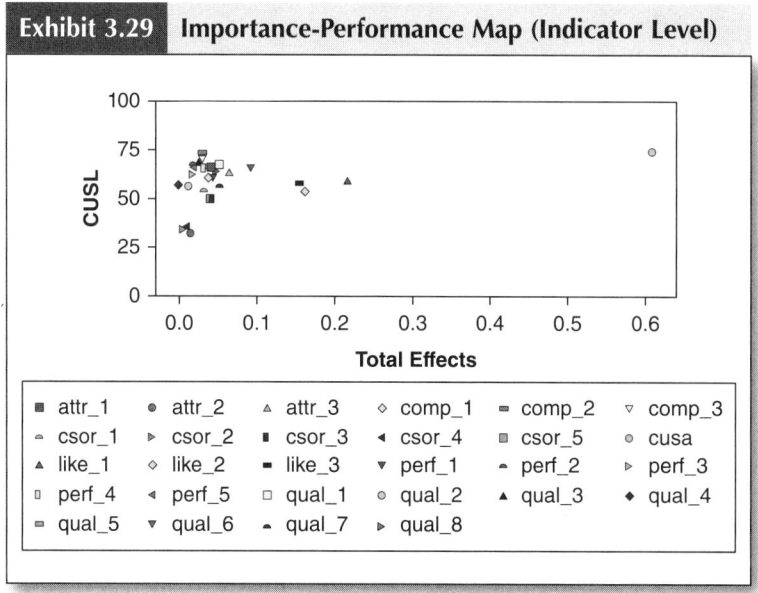

Exhibit 3.29 **Importance-Performance Map (Indicator Level)**

■ attr_1	● attr_2	▲ attr_3	◇ comp_1	▰ comp_2	▽ comp_3
◠ csor_1	▹ csor_2	▪ csor_3	◂ csor_4	▨ csor_5	◎ cusa
▲ like_1	◇ like_2	▪ like_3	▼ perf_1	▴ perf_2	▹ perf_3
▫ perf_4	◂ perf_5	▫ qual_1	○ qual_2	▲ qual_3	◆ qual_4
▱ qual_5	▼ qual_6	▴ qual_7	▹ qual_8		

outputs for *QUAL*'s indicators, which enable producing the importance-performance map as shown in Exhibit 3.31. The IPMA reveals that the indicator *qual_6* (i.e., [The company] is a reliable partner for customers), has a particularly high importance for improving customer loyalty. Similarly, *qual_7* (i.e., [The company] is a trustworthy company) has a relatively high importance but also shows a particularly low performance. These results suggest that companies should aim at stressing their reliability and trustworthiness when aiming to improve their customers' loyalty.

The results of the IPMA provide the empirical foundations for a better discussion on how to improve the performance of a key target construct. In particular, IPMA identifies specific indicators with high importance but relatively low performance that then become the focal areas for managing activities to achieve performance improvements. Subsequent qualitative and quantitative analyses may include additional important factors such as costs and time, and as a result may entail a revised prioritization of activities. For example, an area with high priority for performance improvements, as revealed by the IPMA, may drop into a lower priority rank, if its performance improvements are very costly and require much time. As a result of these analyses, you obtain a comprehensive picture of effective

Exhibit 3.30 Data of the Importance-Performance Map for *QUAL*'s Indicators		
Indicators of QUAL	*Importance*	*Performance*
qual_1	0.051	67.539
qual_2	0.010	56.202
qual_3	0.025	68.023
qual_4	−0.001	56.880
qual_5	0.039	66.860
qual_6	0.091	65.407
qual_7	0.052	56.638
qual_8	0.047	63.953
Mean Value	0.039	62.688

Exhibit 3.31	Importance-Performance Map for *QUAL*'s Indicators

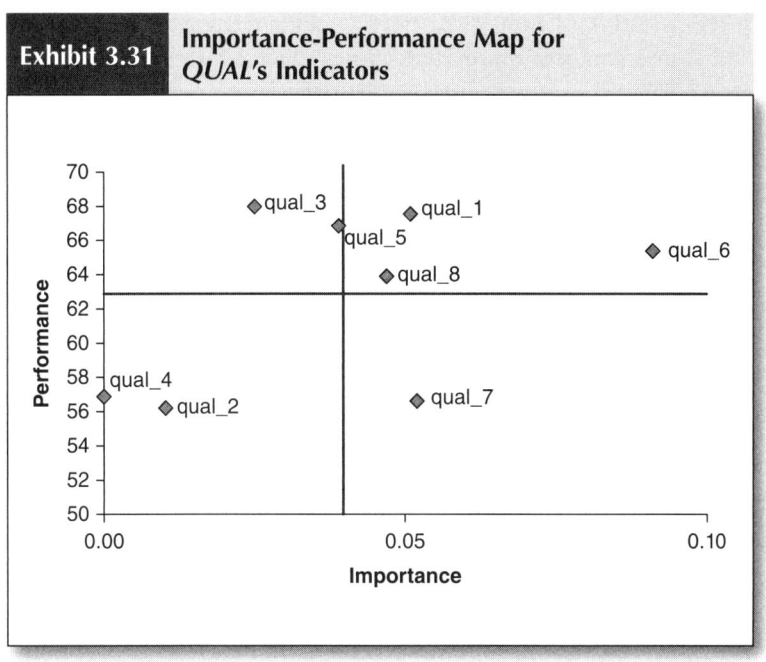

actions for the overall performance improvement opportunities for your key construct of interest (e.g., customer loyalty).

SUMMARY

- **Know how to assess the mode of a measurement model with the confirmatory tetrad analysis in PLS-SEM (CTA-PLS).** CTA-PLS is an approach to empirically evaluate a latent variable's mode of measurement model (i.e., formative or reflective). The method requires, ideally, at least four indicators per measurement model. In case of a reflective measurement model, all its nonredundant tetrads are zero and therefore vanish. However, if at least one of the nonredundant tetrads is significantly different from zero, one should consider rejecting the reflective measurement model and, instead, assume a formative specification. When evaluating the mode of a measurement model, you should primarily rely on theoretical and conceptual considerations along with the empirical evidence provided by CTA-PLS.

- **Understand how to run the CTA-PLS in SmartPLS and how to interpret the results.** Executing the CTA-PLS by using the SmartPLS software requires selecting a large number of bootstrap samples (typically 5,000). Based on the bootstrapping results, SmartPLS returns the bias-corrected and Bonferroni-adjusted confidence intervals for the nonredundant tetrads of all measurement models with four or more indicators. Based on these results, you can decide whether a nonredundant tetrad is statistically significant (i.e., whether it vanishes). Thereby, SmartPLS offers a straightforward means of empirically testing the mode of measurement model.

- **Appreciate the importance-performance map analysis (IPMA).** The purpose of the IPMA is to extend the standard PLS-SEM results reporting of path coefficient estimates by adding a dimension that considers the average values of the latent variable scores. More precisely, the IPMA contrasts the total effects, representing the predecessor constructs' importance in shaping a certain target construct, with their average latent variable scores indicating their performance. The goal is to identify predecessors that have a relatively high importance for the target construct (i.e., those that have a great total effect) but also have a relatively low performance (i.e., low average latent variable scores).

- **Comprehend how to conduct the IPMA using the SmartPLS software.** Before running an IPMA in SmartPLS, researchers need to check several requirements, which relate to the indicator scaling and the indicator weights as obtained from a standard PLS-SEM analysis. The IPMA as such requires selecting a key target construct and specifying the minimum and maximum values of the indicator scales. SmartPLS produces importance-performance maps on the construct level as well as on the indicator level.

REVIEW QUESTIONS

1. What is a tetrad and how would you compute it?

2. What are the key steps of CTA-PLS?

3. How do you decide whether a measurement model is reflective or formative based on the CTA-PLS results?

4. What is the purpose of the IPMA?

5. When looking at the importance-performance map, what does a ceteris paribus interpretation of results mean?

CRITICAL THINKING QUESTIONS

1. Why is measurement model misspecification a critical problem?

2. Use Jarvis et al.'s (2003) qualitative decision criteria to design a reflective and formative approach for measuring customers' satisfaction with their mobile phone service provider.

3. If the CTA-PLS results do not match the assumed measurement model specification, how do you finally decide which kind of measurement mode to use?

4. How can you systematically draw practical implications from IPMA results? When and how would you include the indicator layer?

5. Name an additional (third) interesting dimension of an IPMA that you would consider. Why and how can these additional considerations change your previous prioritization of activities?

KEY TERMS

Alpha inflation

Bias-corrected and Bonferroni-adjusted confidence intervals

Confirmatory tetrad analysis in PLS-SEM (CTA-PLS)

CTA-PLS

Direct effect

Equidistant scale

Forced-choice scale

Formative measurement model

Impact-performance map

Importance

Importance-performance map analysis (IPMA)

Importance-performance matrix

Indirect effect

IPMA

Measurement model misspecification

Metric scale

Multiple testing problem

Nominal scale

Nonredundant tetrads

Ordinal scale

Performance	Rescaling
Priority map analysis	Tetrad (τ)
Reflective measurement model	Total effect
	Vanishing tetrad

SUGGESTED READINGS

Bollen, K. A., & Ting, K.-F. (2000). A tetrad test for causal indicators. *Psychological Methods, 5*, 3–22.

Gudergan, S. P., Ringle, C. M., Wende, S., & Will, A. (2008). Confirmatory tetrad analysis in PLS path modeling. *Journal of Business Research, 61*, 1238–1249.

Höck, C., Ringle, C. M., & Sarstedt, M. (2010). Management of multipurpose stadiums: Importance and performance measurement of service interfaces. *International Journal of Services Technology and Management, 14*, 188–207.

Mikulic, J., Prebežac, D., & Dabic, M. (2016). Importance-performance analysis: Common misuse of a popular technique. *International Journal of Market Research, 58*, 775–778.

Ringle, C. M., & Sarstedt, M. (2016). Gain more insight from your PLS-SEM results: The importance-performance map analysis. *Industrial Management & Data Systems, 116*, 1865–1886.

Schloderer, M. P., Sarstedt, M., & Ringle, C. M. (2014). The relevance of reputation in the nonprofit sector: The moderating effect of sociodemographic characteristics. *International Journal of Nonprofit and Voluntary Sector Marketing, 19*, 110–126.

CHAPTER 4

Modeling Observed Heterogeneity

CHAPTER PREVIEW

Relationships in PLS path models imply that exogenous latent variables explain endogenous latent variables without any systematic influences of other variables. In many instances, however, this assumption does not hold. For example, respondents are likely to be heterogeneous in their perceptions and evaluations of latent variables, yielding significant differences in, for example, path coefficients across two or more groups of respondents (e.g., females vs. males). Recognizing that heterogeneous data structures are often present, researchers are increasingly interested in identifying and understanding such differences. In fact, failure to consider heterogeneity can be a threat to the validity of partial least squares structural equation modeling (PLS-SEM) results (Becker, Rai, Ringle, & Völckner, 2013; Hair, Sarstedt, Ringle, & Mena, 2012).

As a solution, we learn about different concepts that enable researchers to model heterogeneous data. This chapter first provides an overview of observed and unobserved heterogeneity, showing how disregarding heterogeneous data structures can provoke biased results. Next, we discuss measurement invariance, which is a primary concern before comparing groups of data. By establishing measurement invariance, researchers can be confident that group differences in model estimates result from neither the distinctive content and/or meanings of the latent variables across groups nor the measurement scale. To assess measurement invariance in a PLS-SEM context, researchers can use the measurement invariance of the composite models (MICOM) procedure. Next, we introduce different types of multigroup analyses that are used to compare parameters (usually path coefficients) between two or more groups of data. A PLS-SEM multigroup analysis is typically applied when researchers want to explore differences that can be traced back to observable characteristics such as gender or country of origin. In this situation, it is assumed that there is a categorical moderator variable (e.g., gender) that influences the relationships in the PLS path model. Yet, while categorical data allow easy specification of groups, other data can also provide a basis to specify groups. The aim of multigroup analysis, therefore, is to disclose the effect of this categorical moderator variable. The chapter closes with an illustration of measurement model invariance assessment using the MICOM procedure and a PLS multigroup analysis.

OBSERVED AND UNOBSERVED HETEROGENEITY

Applications of PLS-SEM usually analyze the full set of data, implicitly assuming that the data stem from a homogeneous population. This assumption of relatively homogeneous data characteristics is often unrealistic. Individuals (e.g., in their behavior), corporations (e.g., in their structure), or environments (e.g., in their dynamism) are frequently different, and pooling data across observations is likely to produce misleading results. Failure to consider such heterogeneity can be a threat to the validity of PLS-SEM results, since it can lead to incorrect conclusions (Sarstedt & Ringle, 2010). Becker, Rai, Ringle, and Völckner (2013) address this problem in detail and provide examples of situations with significantly different positive and negative path coefficients, which show a nonsignificant value close to 0 on the aggregate data level. Concluding that no relationship exists between the constructs would be invalid and misleading.

The model shown in Exhibit 4.1, in which customer satisfaction with a product (Y_3) depends on the two perceptual dimensions— satisfaction with the quality (Y_1) and satisfaction with the price (Y_2)— illustrates the problems stemming from failure to treat heterogeneity in the context of PLS-SEM. Suppose there are two segments of similar sample sizes. Group 1 comprises male customers, whereas Group 2 consists of female customers. Both groups differ regarding their price consciousness as indicated by the different segment-specific path coefficients. More precisely, the effect of perceived quality (Y_1) on customer satisfaction (Y_3) is much stronger among males ($p_1^{(1)} = 0.50$; the superscript in parentheses indicates the group) than among females ($p_1^{(2)} = 0.10$). In contrast, perceived price (Y_2) has a somewhat stronger influence on customer satisfaction (Y_3) among females ($p_2^{(2)} = 0.35$) than among males ($p_2^{(1)} = 0.25$). In this example, heterogeneity reflects the results of one group of customers (females) being more price sensitive but less quality sensitive, whereas it is the opposite for the other group (males). From a technical perspective, there is a categorical moderator variable *gender* that splits the data set into two customer groups and thus requires estimating two separate models, as indicated in Exhibit 4.1. Importantly, if we fail to recognize the heterogeneity between the groups and analyze the model using the full set of data, the path coefficient estimates offer an incomplete picture of the model relationships. That is, both estimates would equal approximately 0.30 when using the full set of data, thus leading the researcher to conclude that price and quality are equally important for males and females, although they are not. Similarly, group-related differences in model estimates can cancel each other out, yielding nonsignificant effects when analyzing the data on the aggregate level (for an example, see Sarstedt, Schwaiger, & Ringle, 2009). Consequently, it is important to identify, assess, and, if present, treat heterogeneity in the data.

Heterogeneity in data can be observed or unobserved. When differences between two or more groups of data relate to observable characteristics, such as gender, age, or country of origin, this is **observed heterogeneity**. Researchers can use these observable characteristics to partition the data into separate groups of observations and carry out group-specific PLS-SEM analyses, as illustrated in Exhibit 4.1

Exhibit 4.1	Heterogeneity in PLS Path Models

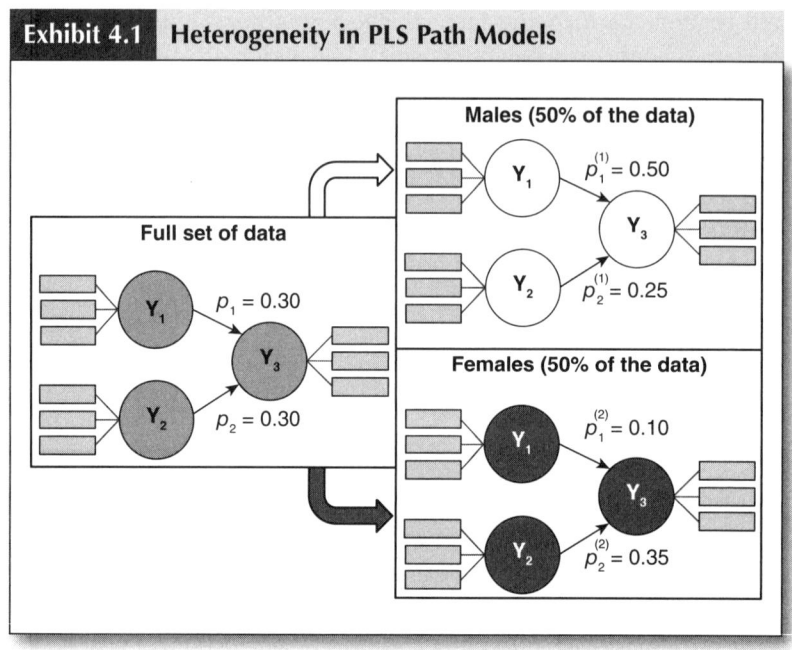

with regard to customers' gender. On the contrary, **unobserved heterogeneity** implies that differences between two or more groups of data do not emerge a priori from a specific observable characteristic or combinations of several characteristics but becomes apparent in differences in structural path coefficients.

In an attempt to account for unobserved heterogeneity with regard to endogenous but also exogenous variables, researchers have routinely used **cluster analysis** techniques, such as **k-means** (Hair, Black, Babin, & Anderson, 2010; Sarstedt & Mooi, 2014) on the indicator data, or latent variable scores derived from a preceding analysis of the entire data set. The partition that this analysis produces is then used as input for group-specific PLS-SEM estimations. While easy to apply, such an approach is conceptually flawed because traditional clustering techniques ignore the path model relationships that researchers specified prior to the analysis. But it is exactly these relationships that are likely responsible for some of the group differences. Therefore, it is not surprising that prior research has shown that traditional clustering approaches perform very poorly in identifying group differences in PLS-SEM (e.g., Sarstedt & Ringle, 2010).

Recognizing the limitations of sequential approaches, methodological research in PLS-SEM has proposed several specific methods to identify and treat unobserved heterogeneity, commonly referred to as **latent class techniques**. These techniques have a long tradition in covariance-based SEM (CB-SEM) research (e.g., Jedidi, Jagpal, & DeSarbo, 1997; Masyn, 2013; Muthén, 1989) and have proven very useful in identifying unobserved heterogeneity and partitioning the data accordingly. The resulting partition can then be analyzed for significant differences using multigroup analysis approaches. Alternatively, latent class techniques may ascertain that unobserved heterogeneity does not influence the results, supporting an analysis of a single model based on the aggregate-level data.

The different techniques for identifying unobserved heterogeneity in PLS-SEM include genetic algorithms (Ringle, Sarstedt, & Schlittgen, 2014; Ringle, Sarstedt, Schlittgen, & Taylor, 2013), weighted least squares (Schlittgen, Ringle, Sarstedt, & Becker, 2016), and several other approaches (e.g., Becker, Rai, Ringle, & Völckner, 2013; Esposito Vinzi, Trinchera, Squillacciotti, & Tenenhaus, 2008). See Hair, Sarstedt, Matthews, and Ringle (2016) for a review. In Chapter 5, we introduce two latent class procedures, **finite mixture partial least squares** (**FIMIX-PLS**) and **prediction-oriented segmentation in PLS-SEM** (**PLS-POS**), whose joint use allows for reliably identifying and treating unobserved heterogeneity in PLS path models.

TESTING MEASUREMENT MODEL INVARIANCE

A primary concern before comparing group-specific parameter estimates for significant differences using a multigroup analysis is ensuring **measurement invariance**, also referred to as **measurement equivalence**. By establishing measurement invariance, researchers can be confident that group differences in model estimates do not result from the distinctive content and/or meanings of the latent variables across groups. Variations in the structural relationships between latent variables could stem from different meanings the groups' respondents attribute to the phenomena being measured, rather than the true differences in the structural relationships. Reasons for such differences may stem from, for example, (a) respondents embracing different cultural values who interpret a given measure in a conceptually different manner; (b) gender, ethnic, or other individual differences that entail responding to instruments in systematically different ways; and (c) respondents who use the available options on a scale differently (e.g., tendency to choose or not to choose the extremes). Hult et al. (2008, p. 1028) describe these concerns and conclude that "failure to establish data equivalence is a potential source of measurement error" (i.e., discrepancies between what is intended to be measured and what is actually measured). When measurement invariance is not present, it can reduce the power of statistical tests, influence the precision of estimators, and provide misleading results. In short, when measurement invariance is not demonstrated, any conclusions about model relationships are questionable. Hence, multigroup comparisons require establishing measurement invariance to ensure the validity of outcomes and conclusions.

Researchers have suggested a variety of methods to assess measurement invariance for CB-SEM. Multigroup confirmatory factor analysis based on the guidelines of Steenkamp and Baumgartner (1998) and Vandenberg and Lance (2000) is by far the most common approach to invariance assessment. However, the well-established measurement invariance techniques used to assess CB-SEM's **common factor models** and related extensions to formative measurement models (Diamantopoulos & Papadopoulos, 2010) cannot be readily transferred to PLS-SEM's composite models. For this reason, Henseler, Ringle, and Sarstedt (2016) developed the **measurement invariance of composite models (MICOM) procedure**. The MICOM procedure builds on the scores of the latent variables. In PLS-SEM, these latent variables are represented as composites, that is, linear combinations of indicators and the indicator weights as estimated by the PLS-SEM algorithm. Therefore, in the description of the MICOM procedure, we talk about composites when referring to the entities (scores) the PLS-SEM algorithm uses to represent the latent variables as specified by the researcher.

The MICOM procedure involves three steps: (1) **configural invariance**, (2) **compositional invariance**, and (3) **equality of composite mean values and variances**. The three steps are hierarchically interrelated, as displayed in Exhibit 4.2. This means that configural invariance is a precondition for compositional invariance, which is again a precondition for meaningfully assessing the equality of composite mean values and variances.

If configural invariance (Step 1) and compositional invariance (Step 2) are established, **partial measurement invariance** is confirmed. When partial measurement invariance is confirmed for all latent

Exhibit 4.2 The MICOM Procedure

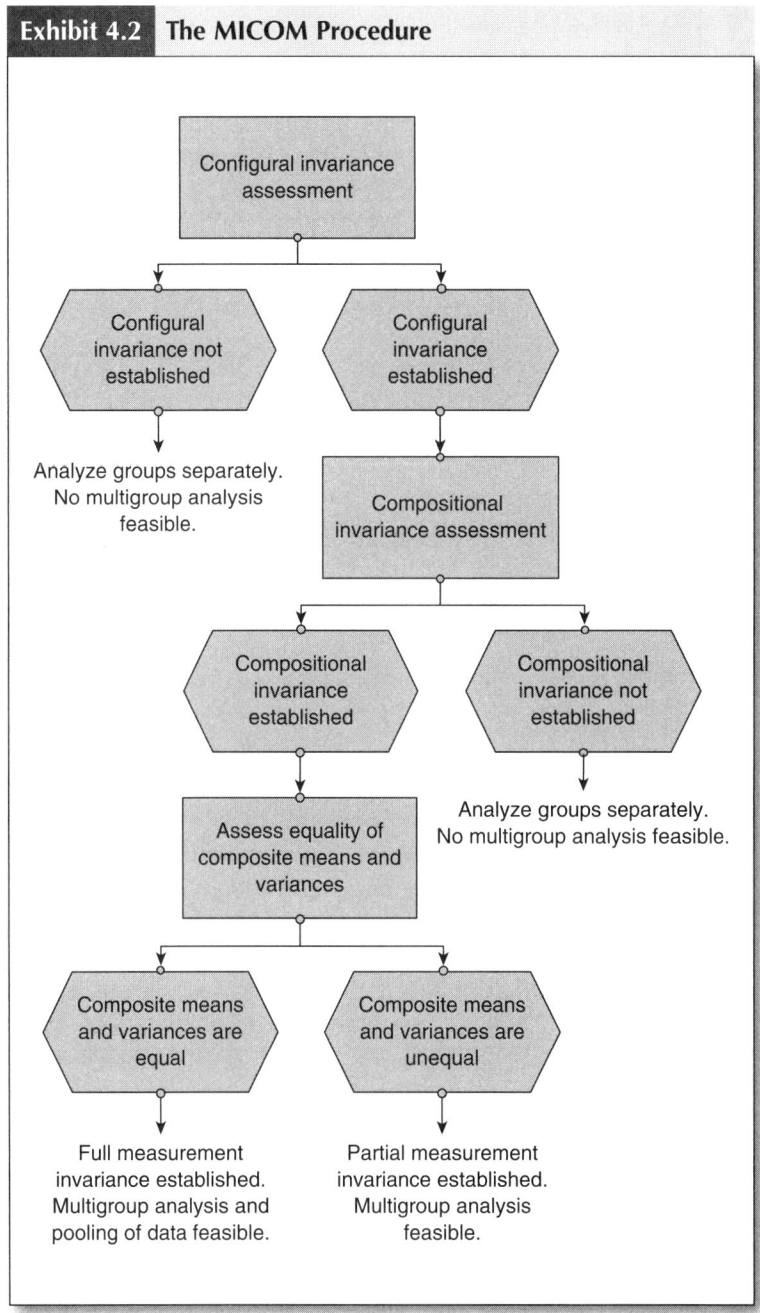

variables in the PLS path model, researchers can compare the path coefficients by means of a multigroup analysis. If partial measurement invariance is established and, additionally, the composites have equal mean values and variances across the groups, **full measurement invariance** is confirmed. From a measurement model perspective, researchers can then pool the data of the different groups, if they come from separate groups initially, and benefit from the increase in statistical power. However, full measurement invariance does not imply that there are no differences in the structural models (i.e., the path coefficients). The latter need to be tested by means of a multigroup analysis. Only if the multigroup analysis indicates that the structural models are also invariant (i.e., there are no significant differences in the path coefficients) can researchers pool the data and exclusively focus on the aggregate-level analysis. In the following sections, we discuss each step in greater detail (for an application of MICOM on data from five countries, see Schlägel & Sarstedt, 2016).

Step 1: Configural Invariance

Step 1 addresses the establishment of configural invariance to ensure that each latent variable in the PLS path model has been specified equally for all the groups. Configural variance exists when constructs are equally parameterized and estimated across groups. An initial qualitative assessment of the latent variables' specification across all the groups must ensure that the following three requirements have been met:

1. *Identical indicators per measurement model.* Each measurement model must employ the same indicators and scale across the groups. Checking whether exactly the same indicators apply to all the groups seems rather simple. However, when conducting surveys using different languages, the application of good empirical research practices (e.g., translation and back-translation) is of the utmost importance to establish the indicators' equivalence. In this context, an assessment of face and/or expert validity can help verify whether the researcher(s) used the same set of indicators across the groups.

2. *Identical data treatment.* The indicators' data treatment must be identical across all the groups, which includes the coding (e.g., dummy coding), reverse coding, and other

forms of recoding as well as the data handling (e.g., standardization or missing value treatment). Outliers should be detected and treated similarly.

3. *Identical algorithm settings or optimization criteria.* Variance-based model estimation methods such as PLS consist of many variants with different target functions and algorithm settings (e.g., choice of initial outer weights and the inner model weighting scheme; Hair, Ringle, & Sarstedt, 2011; Henseler, Ringle, & Sinkovics, 2009). Researchers must ensure that differences in the group-specific model estimations do not result from dissimilar algorithm settings.

Configural invariance is a necessary but not sufficient condition for drawing valid conclusions from multigroup analyses. Researchers also must ensure that differences in path coefficients do not result from differences in the way a latent variable is formed across the groups. The next step, compositional invariance, focuses on this aspect.

Step 2: Compositional Invariance

Compositional invariance exists when the composite scores are the same across the groups, despite possible differences in the group-specific weights used to compute the scores. Step 2 of the MICOM procedure applies a statistical test to assess whether the composite scores differ significantly across the groups. For this purpose, the MICOM procedure examines c, which is the correlation between the composite scores $Y^{(1)}$ and $Y^{(2)}$:

$$c = \text{cor}\left(Y^{(1)}, Y^{(2)}\right).$$

Compositional invariance requires that c equals 1. Technically, the procedure tests the null hypothesis that c is 1. In order to establish compositional invariance, we must not reject this null hypothesis. That is, if the test yields a p value larger than 0.05 (in case we assume a significance level of 5%), we can assume compositional invariance. On the contrary, if we reject the null hypothesis, we cannot establish compositional invariance for the specific construct under consideration. Note that this test does not work for single-item constructs since their (single) outer relationship always is 1. Hence, the latent variable scores $Y^{(1)}$ and $Y^{(2)}$ of a single item construct are identical entailing a c value of 1.

For hypothesis testing, the MICOM procedure draws on the concept of **permutation** (Fisher, 1935). Similar to bootstrapping, permutation tests generate a reference distribution from the actual data. However, instead of sampling observations from the original data set with replacement—as it is the case in bootstrapping—permutation tests randomly exchange observations between the groups multiple times. Permutation testing has routinely been used in a multitude of contexts as it provides an efficient approach to nonparametric testing, also when the sample size is small (e.g., Ernst, 2004; Good, 2000).

The compositional invariance assessment in MICOM follows a five-step approach, which Exhibit 4.3 illustrates:

1. Group-specific estimation of the PLS path models to obtain composite scores of group 1 with $n^{(1)}$ observations and group 2 with $n^{(2)}$ observations. Note that in the example of two groups the sum of $n^{(1)}$ and $n^{(2)}$ equals n, which is the total number of observations in the full set of data.

2. Computation of the correlations c between the composite scores of groups 1 and 2.

3. Random permutation of the data, meaning that the observations are randomly assigned to the groups. In formal terms, $n^{(1)}$ observations are randomly drawn without replacement from the aggregate data set and assigned to group 1. The remaining $n^{(2)}$ observations are assigned to group 2. Researchers should use a minimum of 1,000 permutations.

4. For each permutation run u computations of the correlation c_u between the composite scores of group 1 and group 2.

5. For hypothesis testing, the procedure sorts the permutations' correlation results c_u in descending order. The cutoff value is the 95% point (or, in our example, the 950th of 1,000 permutations in the sorted list) and the corresponding correlation value. If c is smaller than that value, it falls out of the 95% permutation-based **confidence interval** and is significantly different from 1 (at the $p < 0.05$ level), which entails a rejection of the null hypothesis that c equals 1.

Exhibit 4.3 Compositional Invariance Testing in the MICOM Procedure

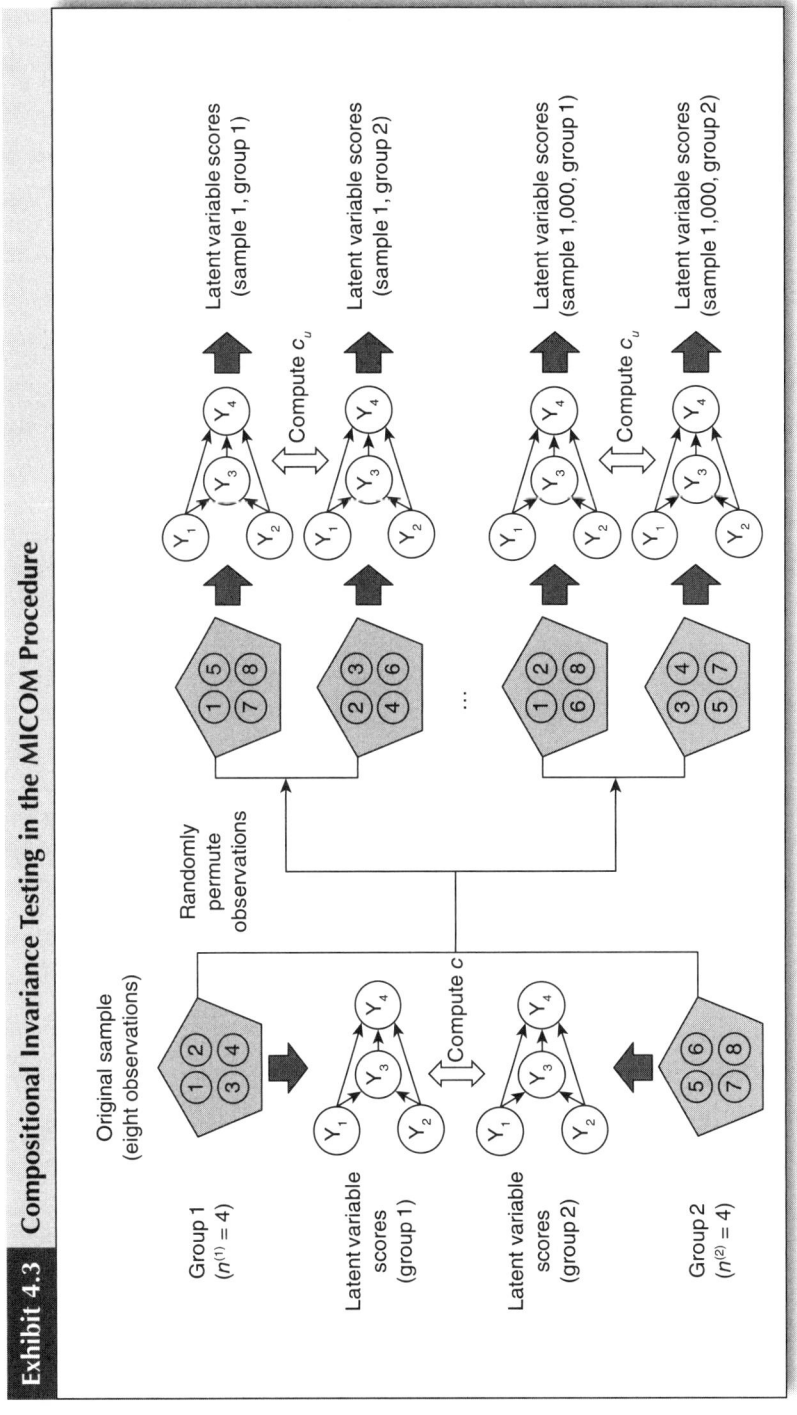

Consequently, the composite scores are significantly different, and compositional invariance has not been established. In the opposite situation, if c does not fall into the 5% extreme tail, but into the 95% confidence interval, the composite scores are not significantly different, which substantiates compositional invariance. Based on the permutation results, it is also possible to return the p value of c. In this analysis, the p value represents the percentage of the permutations' correlation c_u results that have a lower value than c. A p value above 0.05 indicates that c is not significantly different from 1, which means that compositional invariance has been established.

When configural invariance exists and compositional invariance is established for all latent variables in the PLS path model, there is partial measurement invariance. In this case, researchers can compare the path coefficients by means of a multigroup analysis. If partial measurement invariance cannot be established for a latent variable, then group-specific comparisons using a multigroup analysis on any relationships involving this latent variable are not feasible. In this case, researchers may refrain from running a multigroup analysis altogether and analyze each group separately without attempting to compare the results for the two groups. Alternatively, one can eliminate the latent variables that did not achieve compositional invariance, provided theory supports this step. Importantly, when making any changes to the model, MICOM needs to be rerun.

Step 3: Equality of Composite Mean Values and Variances

If the results of Step 2 support measurement invariance, the assessment should continue with the equality assessment of the composites' mean values and variances.

This final step of the MICOM procedure first requires estimating the PLS path model using the pooled (i.e., aggregate) data. We then examine whether the mean values and variances between the composite scores of the first group and the composite scores of the second group differ regarding their means and variances. Note that, different from Step 2, we do not use the composite scores computed by means of separate, group-specific PLS-SEM estimations but, instead, use the entire data set and assign the scores to each group ex post. For the analysis of the mean values' equivalence, the null hypothesis is

$$H_0 : \overline{Y}_{\text{pooled}}^{(1)} - \overline{Y}_{\text{pooled}}^{(2)} = 0.$$

Here, the index "pooled" indicates that the composite scores of a certain group stem from the analysis of the pooled data.

Analyzing the equivalence of the variances requires determining the logarithm of the variance ratio of the composite scores of both groups. If the logarithm of this ratio is statistically not different from 0, we conclude that the variances are equal across groups. The corresponding null hypothesis is

$$H_0 : \log\left(\operatorname{var}\left(Y_{\text{pooled}}^{(1)}\right) / \operatorname{var}\left(Y_{\text{pooled}}^{(2)}\right)\right) =$$
$$\log\left(\operatorname{var}\left(Y_{\text{pooled}}^{(1)}\right)\right) - \log\left(\operatorname{var}\left(Y_{\text{pooled}}^{(2)}\right)\right) = 0.$$

The testing of these two hypotheses follows a similar approach as in Step 2. MICOM permutes (i.e., rearranges) the observations' group membership many times and generates the empirical distribution of the differences in mean values and logarithms of variances. Full measurement invariance is established when there are no significant differences in mean values and (logarithms of) variances across the groups. This is the case if the permutation-based confidence intervals of the differences in mean values and (logarithms of) variances include the original differences in mean values and variances as obtained by the original model estimation (i.e., without permutation). In contrast, if one of these differences is significant, we must acknowledge that full measurement invariance cannot be established.

In summary, comparing group-specific model relationships for significant differences using a multigroup analysis requires establishing configural (Step 1) and compositional (Step 2) invariance. If these two steps support measurement invariance, partial measurement invariance is established. As the focus is usually on the structural model relationships, we can then run a multigroup analysis on the path coefficients using one of the approaches described in the following sections. However, an exception in this regard is the comparison of interaction terms, created in the course of a moderator analysis (see Hair et al., 2017). The measurement model of an interaction term represents an auxiliary measurement that incorporates the interrelationships between the moderator and the exogenous construct in the path model. This characteristic, however, renders any measurement model assessment of the interaction term and related comparisons across different groups in terms of invariance assessment meaningless. Importantly, if partial measurement invariance cannot be established for a latent variable, group-specific comparisons using a multigroup analysis on any relationships involving this latent variable are not feasible.

To reiterate, researchers may analyze the groups separately, if partial measurement invariance is not apparent, without drawing any conclusions about group-related effects or change the model by eliminating the latent variables that did not achieve compositional invariance. Finally, if partial measurement invariance is confirmed and the composites have equal mean values and variances across the groups, full measurement invariance is confirmed, which additionally supports the pooled data analysis, when considered applicable.

Multigroup Analysis

Path coefficients generated from different samples are almost always numerically different, but the question is whether the differences are statistically significant. **Multigroup analysis** helps to answer this question. Technically, a multigroup analysis tests the null hypotheses H_0 that the path coefficients between two groups (e.g., $p^{(1)}$ in group 1 and $p^{(2)}$ in group 2) are not significantly different (e.g., $p^{(1)} = p^{(2)}$), which amounts to the same as saying that the absolute difference between the path coefficients is 0 (i.e., $H_0 : \left| p^{(1)} - p^{(2)} \right| = 0$). The corresponding alternative hypothesis H_1 is that the path coefficients are different (i.e., $H_0 : p^{(1)} \neq p^{(2)}$ or, put differently, $H_0 : \left| p^{(1)} - p^{(2)} \right| > 0$).

Research has proposed several approaches to multigroup analysis that are illustrated in Exhibit 4.4 (Sarstedt, Henseler, & Ringle, 2011). When comparing two groups of data, researchers need to distinguish between the parametric test and several nonparametric approaches.

The Parametric Test

The **parametric test** (Keil et al., 2000) was the first approach applied in PLS-SEM studies and has been widely adopted because of its ease of implementation. This approach is a modified version of a standard

Exhibit 4.4	Multigroup Analysis Approaches in PLS-SEM

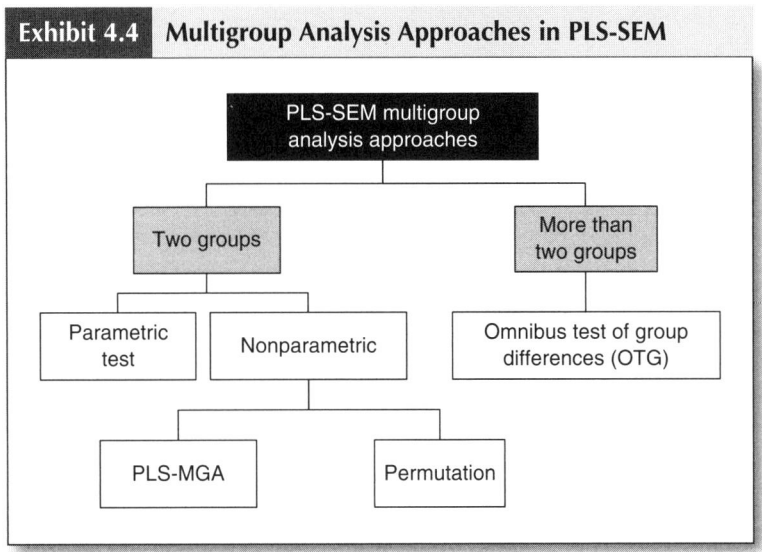

two independent samples t test, which relies on standard errors derived from **bootstrapping**. As with the standard t test, the parametric approach has two versions (Sarstedt & Mooi, 2014), depending on whether population variances can be assumed to be equal (homoscedastic) or unequal (heteroscedastic). The latter can be tested by means of the Levene's test as described in, for example, Sarstedt and Mooi (2014).

If the variances are equal, the test statistic (i.e., the empirical t value) is computed as follows:

$$t = \frac{p^{(1)} - p^{(2)}}{\sqrt{\frac{\left(n^{(1)} - 1\right)^2}{\left(n^{(1)} + n^{(2)} - 2\right)} \cdot \text{se}(p^{(1)})^2 + \frac{\left(n^{(2)} - 1\right)^2}{\left(n^{(1)} + n^{(2)} - 2\right)} \cdot \text{se}(p^{(2)})^2} \cdot \sqrt{\frac{1}{n^{(1)}} + \frac{1}{n^{(2)}}}}$$

In this formula, $p^{(1)}\left(p^{(2)}\right)$ describes the path coefficient to be compared in group 1 (group 2), whereas $n^{(1)}\left(n^{(2)}\right)$ stands for the number of observations in group 1 (group 2). Finally, $se\left(p^{(1)}\right)\left(se\left(p^{(2)}\right)\right)$ describes the standard errors of the parameter estimates of group 1 (group 2), which can be obtained via bootstrapping. To reject the null hypothesis of equal path coefficients, the empirical t value must be larger than the critical value from a t distribution with $n^{(1)} + n^{(2)} - 2$ degrees of freedom.

On the contrary, if the variances are unequal, researchers have to use a modified version of the Welch-Satterthwaite t test (Satterthwaite, 1946; Welch, 1947), whose test statistic takes the following form:

$$t = \frac{p^{(1)} - p^{(2)}}{\sqrt{\frac{\left(n^{(1)} - 1\right)}{n^{(1)}} \cdot se\left(p^{(1)}\right)^2 + \frac{\left(n^{(2)} - 1\right)}{n^{(2)}} \cdot se\left(p^{(2)}\right)^2}}.$$

This test statistic is also asymptotically t distributed but with the following degrees of freedom (df):

$$df = \left\| \frac{\left(\frac{\left(n^{(1)} - 1\right)}{n^{(1)}} \cdot se\left(p^{(1)}\right)^2 + \frac{\left(n^{(2)} - 1\right)}{n^{(2)}} \cdot se\left(p^{(2)}\right)^2 \right)^2}{\frac{\left(n^{(1)} - 1\right)}{n^{(1)^2}} \cdot se\left(p^{(1)}\right)^4 + \frac{\left(n^{(2)} - 1\right)}{n^{(2)^2}} \cdot se\left(p^{(2)}\right)^4} - 2 \right\|.$$

Prior research suggests that the parametric approach is rather liberal and likely subject to Type I errors (Sarstedt, Henseler, & Ringle, 2011). Furthermore, from a conceptual perspective, the parametric approach is problematic since it relies on distributional assumptions, which are inconsistent with PLS-SEM's nonparametric nature. Against this background, researchers have proposed several nonparametric alternatives to multigroup analysis.

NONPARAMETRIC TESTS

PLS-MGA

Henseler et al. (2009) proposed the nonparametric **PLS-MGA** approach that builds on bootstrapping results of each data group. For a specific relationship in the PLS path model, their approach compares each bootstrap estimate of one group with all other bootstrap estimates of the same parameter in the other group. By counting the number of occurrences where the bootstrap estimate of the first group is larger than those of the second group, the approach derives a p value for a one-tailed test. This approach can be described in formal terms as follows:

$$p\left(p^{(1)} \ge p^{(2)} \middle| \beta^{(1)} < \beta^{(2)}\right) =$$

$$1 - \frac{1}{B^2} \sum_j \sum_{\tilde{j}} \Theta\left(\left(p_j^{(1)} + p^{(1)} - \overline{p}^{(1)}\right) - \left(p_{\tilde{j}}^{(2)} + p^{(2)} - \overline{p}^{(2)}\right)\right).$$

With regard to the $\beta^{(1)}$ and $\beta^{(2)}$ true population parameters of each data group, this approach determines the conditional probability $p\left(p^{(1)} \ge p^{(2)} \middle| \beta^{(1)} < \beta^{(2)}\right)$ that the $p^{(1)}$ estimation of a certain path coefficient in group 1 is larger than or equal to its estimated $p^{(2)}$ coefficient in group 2, whereby the superscript in parentheses marks the respective group. By using the PLS-MGA, a researcher would like to ensure that this conditional probability is below a specified α-level before concluding that $\beta^{(1)}$ is greater than $\beta^{(2)}$. In the above equation, B is the number of bootstrap runs, $p_j^{(1)}\left(j=1,\cdots,B\right)$ and $p_{\tilde{j}}^{(2)}\left(\tilde{j} = 1,\cdots, B\right)$ are the group-specific bootstrap results of the path coefficient, $\overline{p}^{(1)}$ and $\overline{p}^{(2)}$ denote the mean values of the group-specific bootstrap samples, and Θ (i.e., the Greek letter *theta*) stands for the unit step function, which has a value of 1 if its argument exceeds 0, otherwise 0. Note that in accordance with Henseler et al. (2009), all group-specific bootstrap results need to be corrected by the difference between the original group-specific parameter estimate and the average over the group-specific bootstrap values.

To illustrate the working principle of the PLS-MGA, consider a simple model with two constructs, estimated in two groups yielding the path coefficients $p^{(1)} = 0.336$ and $p^{(2)} = 0.501$. We want to test the hypothesis that the path coefficient is larger in the first group compared to the second group. To test this hypothesis, we draw 10 **bootstrap samples** for $p^{(1)}$ and $p^{(2)}$ and contrast the estimates as shown in Exhibit 4.5. The first column represents the 10 bootstrap samples of the path coefficient in the first group, whereas the first row represents the 10 bootstrap samples in the second group. We now compare each bootstrap estimate of $p^{(1)}$ (e.g., the first bootstrap sample's estimate 0.357) with each bootstrap estimate of $p^{(2)}$ (i.e., 0.494, 0.423, ..., 0.538) and count the number of cases where

$p^{(1)} \geq p^{(2)}$, indicated by an X in Exhibit 4.5. Dividing this number (i.e., 11) by the total number of comparisons (i.e., 100) yields the p value, which is 0.11 in our case. Therefore, we cannot conclude that the path coefficient p is significantly larger in the first group compared to the second group.

PLS-MGA involves a great number of comparisons of bootstrap estimates. For an initial analysis of group differences, we suggest drawing on 500 bootstrap samples, using the algorithm settings (e.g., no sign change option) as recommended in Hair et al. (2017). For the final analysis, use 5,000 bootstrap samples, which results in 25,000,000 comparisons for each parameter.

Finally, it is important to emphasize that the PLS-MGA approach allows for testing only one-sided hypotheses. Specifically, the SmartPLS software always tests the hypothesis that $p^{(1)}$ is larger than $p^{(2)}$. In case you want to test the opposite direction (i.e., $p^{(2)} > p^{(1)}$), you therefore need to subtract the resulting p value from 1 to obtain the actual p value for our hypothesis. Using the PLS-MGA approach to test two-sided hypotheses is not possible without limitations as the bootstrap-based distribution is not necessarily symmetric. This characteristic clearly limits its applicability as researchers routinely draw on two-tailed tests, despite frequent concerns (e.g., Cho & Abe, 2012).

Permutation Test

The **permutation test** was originally developed by Chin (2003) and further substantiated by Chin and Dibbern (2010). As its name implies and analogous to its role in Step 2 of the MICOM procedure, the permutation test randomly exchanges observations between the data groups and re-estimates the model for each permutation. Computing the differences between the group-specific path coefficients per permutation enables testing whether these also differ in the population.

The permutation test follows a six-step process:

1. Carry out group-specific estimations of the PLS path models to obtain path coefficient estimates for groups 1 and 2.

2. Compute the difference between the group-specific path coefficient estimates: $d = p^{(1)} - p^{(2)}$.

Exhibit 4.5   Data Matrix for PLS-MGA

Bootstrap samples $p^{(2)}$	Bootstrap samples $p^{(1)}$									
	0.357	0.226	0.318	0.281	0.372	0.318	0.296	0.308	0.415	0.272
0.494										
0.423										
0.324	X				X				X	
0.591										
0.698										
0.291	X		X		X	X	X	X	X	
0.509										
0.400									X	
0.526										
0.538										

Note: X indicates situations when $p^{(1)} \geq p^{(2)}$.

3. Produce random permutation of the data, meaning that the observations are randomly assigned to the groups. In formal terms, $n^{(1)}$ observations are drawn without replacement from the aggregate data set and assigned to group 1. The remaining $n^{(2)}$ observations are assigned to group 2. Researchers should use a minimum number of 1,000 permutations.

4. Conduct group-specific estimations of the PLS path models for each permutation run u. For example, with 1,000 permutations, we obtain 1,000 model estimates for group 1 (i.e., $p_u^{(1)}$) and for group 2 (i.e., $p_u^{(2)}$).

5. Compute the differences in the permutation run-specific path coefficient estimates: $d_u = p_u^{(1)} - p_u^{(2)}$.

6. Create the two-tailed 95% permutation-based confidence interval. More specifically, sort the resulting d_u values in ascending order and determine the two values that separate (1) the 2.5% lowest from the 97.5% highest values (lower boundary) and (2) the 97.5% lowest from the 2.5% highest values (upper boundary). If the original difference d of the group-specific path coefficient estimates does not fall into the confidence interval, it is statistically significant. In addition, based on the permutation results, SmartPLS provides the p value of the difference d. A p value smaller than 0.05 suggests that the difference d between the group-specific path coefficients does not fall into the 95% permutation-based confidence interval and, thus, is statistically significant.

Permutation tests in general have been shown to perform very well across a broad range of conditions (e.g., Ernst, 2004; Good, 2000). Most importantly, permutation tests reliably control for Type I errors when the assignment of observations occurs randomly, as is the case in the permutation test's application in PLS-SEM. Correspondingly, prior research indicates that the permutation test performs more conservatively than the parametric test in terms of rendering differences significant (Sarstedt, Henseler, & Ringle, 2011). Against this background and because of the test's nonparametric character, we generally recommend using the permutation test. However, large differences in the group-specific sample sizes have adverse consequences for the permutation test's performance. Therefore, when one group's sample is more than double the size of the other group's, researchers should draw on the PLS-MGA approach (in case of testing a one-sided hypothesis) or the parametric approach. As an alternative approach, researchers can randomly withdraw another sample from the large group that is comparable in size to the smaller group and compare the two samples using the permutation test.

Comparing More Than Two Groups

Standard approaches to multigroup comparison in PLS-SEM have in common that they test the difference in the parameters between two groups. However, researchers frequently encounter situations in which they would like to compare a parameter (e.g., a path coefficient) across more than two groups. In PLS-SEM, the complexity of this analysis—determined by the number of group specific path coefficient comparisons—increases with the path model complexity (i.e., the number of relationships) and the number of groups. For example, a model with 10 path relationships in the structural model and five groups requires 100 comparisons of group-specific path coefficients. If no significant differences exist across groups, the likelihood to find at least one significant difference is larger than 95% when conducting the 100 comparisons. Not only do such complex analyses become increasingly time-consuming, but there is a more severe problem associated with this approach, called **alpha inflation** (also referred to as **multiple testing problem**). This refers to the fact that the more tests you conduct at a certain significance level, the more likely you are to claim a significant result when this is not so (i.e., a Type I error). For example, assuming a significance level of 5% and making all possible pairwise comparisons as specified above, the overall probability of a Type I error (also referred to as the **familywise error rate**) is not 5% but 99.41% (Sarstedt & Mooi, 2014). In other words, when conducting all possible pairwise group comparisons, the familywise error rate quickly increases beyond the acceptable Type I error level (i.e., the acceptance of false positives) of, for example, 5%.

Against this background, when comparing more than two groups, two issues warrant attention. First, we need to test whether the path coefficient differs from the others in at least one of the groups (i.e., testing for the overall difference). For this purpose, we suggest using a permutation-based analysis of variance (ANOVA), which maintains the familywise error rate, does not rely on distributional assumptions, and exhibits an acceptable level of statistical power. In case we find a significant difference, the second step is to assess and determine between which groups the path coefficient differs (i.e., pairwise comparisons) by using the standard procedures (e.g., the permutation test). However, the results become adjusted to account for the multiple testing problem.

In the following section, we explain these two steps in more detail. As a note of caution, it is important to keep in mind that the more you search, the more likely it is that you will find something. For this reason, you should generally focus on revealing substantial differences across groups in PLS-SEM.

Testing for the Overall Difference

The standard approach to testing for the overall difference in a parameter across multiple groups involves running an ANOVA *F* test. However, in light of the test's parametric character and its asymptotic properties when used on bootstrapping results, comparing the different groups by means of an ANOVA *F* test is not meaningful in a PLS-SEM context.

As a remedy, Sarstedt, Henseler, and Ringle (2011) developed the **omnibus test of group differences** (OTG), which combines the bootstrapping procedure with permutation testing to mimic an overall *F* test. This approach maintains the Type I error level as specified by the researcher (i.e., the familywise error rate) and delivers an acceptable level of statistical power while not relying on distributional assumptions. The OTG approach consists of the following steps:

1. Run the bootstrapping procedure on each group separately to derive reference distributions of the group-specific parameters.

2. Compute the variance ratio. This is the ratio of variance explained by the grouping variable and the overall variance:

$$F_R = \frac{K \cdot B \cdot (1/K - 1) \sum_{i=1}^{K} \left(\overline{p}^{(i)} - \overline{p} \right)^2}{1/(B-1) \sum_{i=1}^{K} \sum_{j=1}^{B} \left(p_j^{(i)} - \overline{p}^{(i)} \right)^2}$$

In this equation $p_j^{(i)}$ is the parameter estimate from the *j*th bootstrap sample ($j = 1, \ldots, B$) in the *i*th group ($i = 1, \ldots, K$), $\overline{p}^{(i)}$ the average over all bootstrap parameter estimates of group *i*, and $\overline{p}$ the grand mean of the bootstrap parameter estimates across all the groups.

3. Produce permutations of the parameter estimates $p_j^{(i)}$ between the groups, separately for each bootstrap sample. The process starts with the first bootstrap sample by randomly exchanging the parameter estimates $p_1^{(i)}$ between the groups. Next, the same is done for the second and all further bootstrap samples until all parameter estimates have been permuted for all B bootstrap samples. This process yields a total of $(K!)^{B-1}$ different permutations, which is difficult to complete computationally—for example, for $B=5{,}000$ bootstrap samples and $K=3$ groups, $(3!)^{4{,}999} = 9.508 \cdot 10^{3{,}889}$ permutations are required. Therefore, the OTG approach randomly engages in, for example, 5,000 permutations and computes the variance ratio F_{R_u} for each permutation run $l\,(l=1, \ldots, U)$.

4. Compute the p value using the Heaviside step function H. The Heaviside step function is a discontinuous function whose value is 0 for the negative argument and 1 for the positive argument. In the context of the OTG approach, it returns 1 when $F_R - F_{R_l}$ yields a positive value and 0 for a negative value:

$$p = \frac{1}{U} \sum_{l=1}^{U} H\left(F_R - F_{R_l}\right)$$

If the OTG approach indicates a significant effect, we can conclude that at least one group's path coefficient differs significantly from those of the other groups. This does not mean, however, that all the groups' path coefficients differ significantly. To assess which groups differ from each other, we need to apply a pairwise comparisons test.

The OTG approach has not yet been included in SmartPLS. However, the R code of the OTG approach is available in the download section at http://www.pls-sem.com/.

Pairwise Comparisons

The OTG approach allows identifying candidate path coefficients for a pairwise comparison using, for example, the permutation test. Since such tests compare only two groups at a time, you need to run this test multiple times until you have conducted all relevant comparisons. For example, if you like to compare a specific path relationship p across five groups, you must conduct 10 comparisons.

For this example, Exhibit 4.6 shows how to report the comparisons results of a specific path coefficient across multiple groups (e.g., by reporting the difference and their significance in each cell).

As indicated earlier, a primary concern of conducting pairwise comparisons is the alpha inflation, according to which the overall probability of a Type I error (i.e., the familywise error rate) increases exponentially with the number of comparisons. A standard approach for controlling for the familywise error rate is the **Bonferroni correction**. Instead of using a specific significance level in the test decision, the Bonferroni correction tests each individual hypothesis at a significance level of alpha / m. For example, in case of five groups, there would be $m = 10$ comparisons, yielding a significance level of $0.05 / 10 = 0.005$ instead of 0.05. Another popular approach is the **Šidák procedure**, which uses $1 - (1 - alpha)^{1/m}$ in each of the hypothesis tests. Here, in the example of five groups (i.e., 10 comparisons), one would use a significance level of $1 - (1 - 0.05)^{1/10} = 0.005116$ instead of 0.05.

Both approaches counteract the increase in the familywise error rate when performing multiple comparisons across several groups of data. However, the maintenance of the familywise error rate comes to the expense of statistical power. For example, with $m = 10$, for a single comparison to be considered significant, with the Bonferroni correction the effect needs to be 43% larger than that with an alpha level of 5%.

Exhibit 4.6	Pairwise Comparison of a Specific Path Coefficient Across Multiple Groups				
	Group 1	Group 2	Group 3	Group 4	Group 5
Group 1					
Group 2	$p^{(1)} - p^{(2)}$				
Group 3	$p^{(1)} - p^{(3)}$	$p^{(2)} - p^{(3)}$			
Group 4	$p^{(1)} - p^{(4)}$	$p^{(2)} - p^{(4)}$	$p^{(3)} - p^{(4)}$		
Group 5	$p^{(1)} - p^{(5)}$	$p^{(2)} - p^{(5)}$	$p^{(3)} - p^{(5)}$	$p^{(4)} - p^{(5)}$	

Note: p_1 represents the specific path coefficient in the structural model; the comparison is across five groups; the superscript numbers in brackets indicate the group for which the specific coefficient was obtained.

| Exhibit 4.7 | Rules of Thumb for Invariance Assessment and Multigroup Analyses |

Invariance assessment:

- Running a multigroup analysis requires partial measurement invariance, which is given when configural invariance and compositional invariance hold. If partial measurement invariance is confirmed and the composites have equal mean values and variances across the groups, there is full measurement invariance, which supports the pooled data analysis.

- For configural invariance assessment, check whether the measurement models employ the same indicators across the groups in terms of numbers, item content, and coding. Any type of data treatment (e.g., outliers, missing values) must occur identically for the groups. Group-specific model estimations must draw on the same algorithm settings.

- Use 1,000 permutations (or more) for the assessment of compositional invariance and the equality of composite mean values and variances in Steps 2 and 3 of the MICOM procedure.

- Compositional invariance is established when the original correlations between the composite scores of groups 1 and 2 (i.e., the c values) are larger than the 5%-quantile of the empirical distribution of c_u. In exploratory research and when the sample size is small, reverting to the 1%-quantile is more reasonable.

- To test for the equality of the composites' mean values and variances across the groups, check whether differences in means and variances are significant.

Multigroup analyses:

- Use the permutation test to multigroup analysis using 1,000 permutations (or more).

- When one group's sample is more than double the size of the other group's, randomly draw another sample from the large group that is comparable in size to the smaller group, and compare the two samples using the permutation test. Alternatively, use the PLS-MGA approach (in case of testing a one-sided hypothesis) or the parametric approach.

- When using the PLS-MGA approach, use 500 bootstrap samples for the initial assessment and 5,000 bootstrap samples for the final assessment of group differences. Use standard bootstrapping algorithm settings (e.g., no sign change option).

- When using the parametric approach, test for differences in the group variances. In case variances differ significantly, apply the Welch-Satterthwaite t test.

- To analyze more than two groups, use the OTG approach.

Therefore, researchers generally prefer the Šidák procedure over the Bonferroni correction as this approach achieves an acceptable level of statistical power. Research has brought forward a range of alternative (and more complex) approaches for maintaining the familywise error rate (for details, see Hochberg & Tamhane, 2011).

Exhibit 4.7 summarizes the rules of thumb for a combined multigroup analysis and invariance assessment in PLS-SEM.

CASE STUDY ILLUSTRATION—INVARIANCE ASSESSMENT AND MULTIGROUP ANALYSIS

To illustrate the use of multigroup analysis in conjunction with the assessment of measurement model invariance by means of the MICOM procedure, we draw on the **Extended model** in the **Corporate Reputation - Advanced PLS-SEM Book** project. Rather than analyzing the aggregate data set with 344 observations, we are interested in analyzing whether the effects in the model differ significantly for customers with prepaid cell phone plans from those with contract plans. The research study obtained data on customers with a contract plan (*servicetype* = 1; $n^{(1)} = 219$) versus those with a prepaid plan (*servicetype* = 2; $n^{(2)} = 125$), and we will compare those two groups.

When engaging in a multigroup analysis, we need to ensure that the number of observations in each group meets the rules of thumb for minimum sample size requirements. As the maximum number of arrows pointing at a latent variable is eight, we would need at least $8 \cdot 10 = 80$ observations per group, according to the 10 times rule. Following the more rigorous recommendations from a power analysis (Hair et al., 2017), 54 observations per group are needed to detect R^2 values of around 0.25 at a significance level of 5% and a power level of 80%. Therefore, the group-specific sample sizes can be considered sufficiently large.

In the first step, we have to define the grouping variable that SmartPLS uses to split up the data set. To do so, double-click on the data set **Corporate Reputation Data [344 records]** in the **Corporate Reputation - Advanced PLS-SEM Book** project. SmartPLS will open a new tab (Exhibit 4.8), which provides information on the data set and its format (data view).

When opening the data view, three options appear in the menu above the modeling window: **Add Data Group, Generate Data Groups,** and **Clear Data Groups.** To define a new grouping variable,

Exhibit 4.8	**Data View in SmartPLS**

Delimiter:	Semicolon	Encoding:	UTF-8
Value Quote Character:	None	Sample size:	344
Number Format:	US (e.g. 1,000.23)	Indicators:	41
Missing Value Marker:	-99	Missing Values:	11

Indicators: | Indicator Correlations | Raw File

	No.	Missing	Mean	Median	Min	Max	Standard Deviation	Excess Kurtosis	Skewness
serviceprovider	1	0	2.000	2.000	1.000	4.000	1.003	-0.513	0.747
servicetype	2	0	1.637	2.000	1.000	2.000	0.481	-1.684	-0.571
comp_1	3	0	4.648	5.000	1.000	7.000	1.433	-0.324	-0.264
comp_2	4	0	5.424	6.000	1.000	7.000	1.375	-0.616	-0.566
comp_3	5	0	5.221	6.000	1.000	7.000	1.458	-0.188	-0.677
like_1	6	0	4.584	5.000	1.000	7.000	1.547	-0.399	-0.405
like_2	7	0	4.250	4.000	1.000	7.000	1.848	-0.901	-0.312
like_3	8	0	4.480	5.000	1.000	7.000	1.871	-0.941	-0.325
cusl_1	9	3	5.129	5.000	1.000	7.000	1.513	0.268	-0.792
cusl_2	10	4	5.276	6.000	1.000	7.000	1.744	0.040	-0.951
cusl_3	11	3	5.651	6.000	1.000	7.000	1.655	0.930	-1.301

click on **Generate Data Groups** and a new dialogue box will open (Exhibit 4.9). Next to **Name prefix**, you can specify a prefix that SmartPLS uses in the naming of the groups in the results report. In our example, we maintain the default prefix *Group_*. Under **Group Column,** we can specify one or more grouping variables. When left-clicking on the box next to **Group column 0**, a list of all variables in the dataset will appear. Behind each variable, SmartPLS will show the number of unique values. Scroll down and locate the variable *servicetype*, which has two unique values. Left-click on *servicetype* and close the dialogue box by left-clicking on **OK**.

Back in the data view, SmartPLS will show an additional tab labeled **Data Groups**, which offers information on the group-specific sample sizes. You can now close the data view and go back to the modeling window showing the extended corporate reputation model. Before continuing, make sure that the correct data set (i.e., **Corporate Reputation Data [344 records]**) is activated. If the corresponding font does not appear in green, right-click on **Corporate Reputation Data [344 records]** and click on **Select Active Data File**.

To run the MICOM procedure, you need to navigate to **Calculate** → **Permutation**, which you can find at the top of the

---

Exhibit 4.9 | **Generate Data Groups in SmartPLS**

Generate Data Groups

Name prefix: GROUP_

——— **Group Columns** ———

Group column 0: servicetype (2 unique values) ⌄

Group column 1: ⌄

Group column 2: ⌄

——— **Prune groups** ———

Total: 2

Minimum cases: 80

SmartPLS screen. Alternatively, you can left-click on the wheel symbol with the label **Calculate** in the tool bar. A menu with alternative analysis options opens from which you can select running the **Permutation** procedure.

In the dialogue box that follows (Exhibit 4.10), we first need to specify the groups to be compared. To do so click on the menu next to **Group A** and select *GROUP_servicetype(1.0)*. Next, click on the menu next to **Group B** and select *GROUP_servicetype(2.0)*. Specify at least a number of **1,000** permutations and **two-tailed** testing. Because service type is known to affect the way customers relate to experiences, we assume a significance level of **0.05**. Next, click on

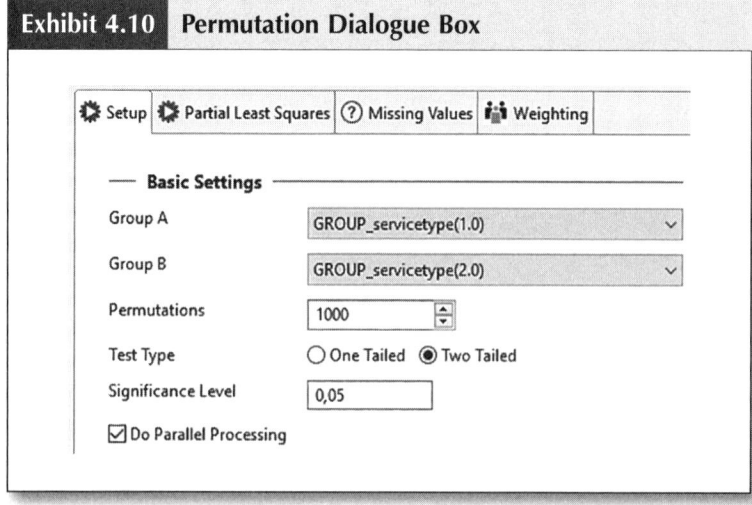

Exhibit 4.10 | **Permutation Dialogue Box**

⚙ Setup | ⚙ Partial Least Squares | ⑦ Missing Values | 👥 Weighting

——— **Basic Settings** ———

Group A   GROUP_servicetype(1.0)   ⌄

Group B   GROUP_servicetype(2.0)   ⌄

Permutations   1000   ⬍

Test Type   ○ One Tailed ● Two Tailed

Significance Level 0,05

☑ Do Parallel Processing

Start Calculation. SmartPLS now estimates the model for each group separately, using the same algorithm settings.

The PLS path models as well as the data treatment used in both groups are identical, which is a necessary requirement for the establishment of configural invariance in Step 1 of the MICOM procedure. Furthermore, as our group-specific model estimations also draw on the identical algorithm settings, configural invariance (Step 1 in MICOM) is established. The following analyses address Steps 2 and 3 of the MICOM procedure. In the results report that opens, go to **Quality Criteria → MICOM**. The first tab labeled **Step 2** (see Exhibit 4.11) shows the results of the compositional invariance assessment. Note that as permutation is a random process, your results will slightly differ from those presented here. Similarly, the results will differ slightly every time you rerun the MICOM procedure.

The column **5%** shows the 5% quantile of the empirical distribution of c_u. Comparing the correlations c between the composite scores of the first and second group (column **Original Correlations**) with the 5% quantile reveals that the quantile is always smaller than (or equal to) the correlation c for all the constructs. This result is also supported by the p values that are higher than 0.05, indicating the correlation is not significantly lower than 1. For example, the original correlation value of *ATTR* is 0.993. This result is within the corresponding permutation-based confidence interval with a lower boundary of 0.945; in accordance, *ATTR*'s p value of 0.673, as displayed in the column **Permutation p-Values**, is considerably larger than 0.05 (Exhibit 4.11). Hence, the original correlation of *ATTR* is not significantly different from 1, which supports the conclusion

---

Exhibit 4.11 **MICOM Step 2 in SmartPLS**

	Step 2	Step 3			
	Original Correlation	Correlation Permutation Mean		5.0%	Permutation p-Values
ATTR	0.993	0.982	0.945		0.673
COMP	0.998	0.999	0.997		0.114
CSOR	0.985	0.961	0.906		0.828
CUSA	1.000	1.000	1.000		0.066
CUSL	1.000	0.999	0.998		0.813
LIKE	1.000	1.000	0.999		0.648
PERF	0.979	0.971	0.933		0.598
QUAL	0.961	0.969	0.940		0.248

that compositional invariance has been established for this construct. Similarly, we substantiate that compositional invariance has been established for all multi-item constructs in the model. Note that we cannot run Step 2 for the single-item construct *CUSA* since its single outer relationship is 1 by design. Thus, we can ignore the *p* value of *CUSA* (Exhibit 4.11), which is not perfectly 0 in this example since the lower boundary value is 0.999999999999998 due to rounding of certain permutation results. To summarize, the results from Step 2 support partial measurement invariance. Thus, we can compare the standardized path coefficients across the groups by means of a multigroup analysis with confidence. To check whether even full measurement invariance holds, click on the tab **Step 3** in the results report (see Exhibit 4.12).

The first two columns in Exhibit 4.12 show the mean differences between the composite scores as resulting from the original model estimation and the permutation procedure, respectively (note that due to space constraints, the screenshot does not show the entire column label). The next two columns show the lower (**2.5%**) and upper (**97.5%**) boundaries of the 95% confidence interval of the scores' mean differences. As we can see, every confidence interval includes the original difference in mean values, indicating that there are no significant differences in the mean values of latent variables across the two groups. For example, for *ATTR*, the original difference in mean values of the latent variable scores is –0.064, which is within the corresponding confidence interval with a lower boundary of –0.225 and an upper boundary of 0.221. The result in the column **Permutation p-Values** further supports this finding for *ATTR* and every other construct in the PLS path model as all the *p* values are considerably larger than 0.05. The next columns show the analogous results for the composite variances. Again, all the confidence intervals include the original value and all the *p* values are clearly larger than 0.05. We therefore conclude that all the composite mean values and variances are equal, providing support for full measurement invariance. Exhibit 4.13 summarizes the results of this MICOM analysis. In light of these results, we continue by examining the results of the multigroup analysis. Our assessment focuses on the permutation test, which we can access by going to **Final Results** → **Path Coefficients** in the SmartPLS permutation output.

Exhibit 4.12   MICOM Step 3 in SmartPLS

Step 2 | Step 3

	Mean - Original...	Mean - Permutation...	2.5%	97.5%	Permutation p-Values	Variance - Original...	Variance - Permutation...	2.5%	97.5%	Permutation p-Values
ATTR	-0.064	-0.004	-0.225	0.221	0.601	0.190	-0.008	-0.297	0.254	0.691
COMP	0.060	-0.001	-0.238	0.236	0.611	0.135	-0.015	-0.298	0.266	0.639
CSOR	0.072	-0.000	-0.216	0.223	0.539	-0.106	-0.012	-0.294	0.262	0.581
CUSA	-0.097	-0.004	-0.226	0.214	0.420	0.230	-0.004	-0.374	0.351	0.613
CUSL	-0.060	-0.001	-0.229	0.219	0.587	-0.070	-0.001	-0.340	0.358	0.721
LIKE	0.054	-0.000	-0.227	0.229	0.653	0.012	-0.010	-0.258	0.248	0.747
PERF	-0.023	-0.003	-0.222	0.221	0.897	0.038	-0.009	-0.306	0.293	0.927
QUAL	-0.010	-0.003	-0.229	0.244	0.949	0.154	-0.010	-0.309	0.271	0.981

Exhibit 4.13	Summary of the MICOM Results

MICOM Step 1

Configural variance established? Yes

MICOM Step 2

Composite	Correlation c	5% quantile of the empirical distribution of c_u	p value	Compositional invariance established?
ATTR	0.993	0.945	0.673	Yes
COMP	0.998	0.997	0.114	Yes
CSOR	0.985	0.906	0.828	Yes
CUSA	1.000	1.000	0.066	Yes
CUSL	1.000	0.998	0.813	Yes
LIKE	1.000	0.999	0.648	Yes
PERF	0.979	0.933	0.598	Yes
QUAL	0.961	0.940	0.248	Yes

MICOM Step 3

Composite	Difference of the composite's mean value (= 0)	95% confidence interval	p value	Equal mean values?
ATTR	−0.064	[−0.225; 0.221]	0.601	Yes
COMP	0.060	[−0.238; 0.236]	0.611	Yes
CSOR	0.072	[−0.216; 0.223]	0.539	Yes
CUSA	−0.097	[−0.226; 0.214]	0.420	Yes
CUSL	−0.060	[−0.229; 0.219]	0.587	Yes
LIKE	−0.054	[−0.227; 0.229]	0.653	Yes
PERF	−0.023	[−0.222; 0.221]	0.897	Yes
QUAL	−0.010	[−0.229; 0.244]	0.949	Yes

Composite	Logarithm of the composite's variances ratio (= 0)	95% confidence interval	p value	Equal variances?
ATTR	0.190	[−0.297; 0.254]	0.691	Yes
COMP	0.135	[−0.298; 0.266]	0.639	Yes
CSOR	−0.106	[−0.294; 0.262]	0.581	Yes
CUSA	0.230	[−0.374; 0.351]	0.613	Yes
CUSL	−0.070	[−0.34; 0.358]	0.721	Yes
LIKE	0.012	[−0.258; 0.248]	0.747	Yes
PERF	0.038	[−0.306; 0.293]	0.927	Yes
QUAL	0.154	[−0.309; 0.271]	0.981	Yes

Exhibit 4.14 **Permutation Test in SmartPLS**

Path Coefficients

Matrix

	Path Coefficients...	Path Coefficients...	Path Coefficients Original Difference ...	Path Coefficients Permutation...	2.5%	97.5%	Permutation p-Values
ATTR -> COMP	0.163	0.049	0.114	−0.002	−0.217	0.228	0.312
ATTR -> LIKE	0.245	0.133	0.112	−0.008	−0.263	0.255	0.352
COMP -> CUSA	0.214	0.108	0.106	−0.003	−0.291	0.276	0.440
COMP -> CUSL	0.136	−0.062	0.197	0.002	−0.214	0.226	0.072
CSOR -> COMP	−0.026	0.091	−0.118	0.001	−0.212	0.225	0.290
CSOR -> LIKE	0.096	0.199	−0.103	0.001	−0.241	0.233	0.380
CUSA -> CUSL	0.599	0.440	0.159	−0.001	−0.166	0.169	0.070
LIKE -> CUSA	0.365	0.476	−0.111	0.003	−0.259	0.251	0.362
LIKE -> CUSL	0.205	0.424	−0.220	−0.002	−0.233	0.222	0.056
PERF -> COMP	0.210	0.305	−0.095	0.007	−0.274	0.287	0.466
PERF -> LIKE	−0.002	0.172	−0.174	0.004	−0.290	0.276	0.224
QUAL -> COMP	0.535	0.418	0.117	0.001	−0.286	0.267	0.418
QUAL -> LIKE	0.501	0.336	0.165	0.008	−0.246	0.271	0.228

The first two columns in Exhibit 4.14 show the original path coefficients in group 1 and group 2, respectively, followed by their differences in the original data set and the permutation testing, respectively (note that due to space constraints, the screenshot does not show the entire column label). As can be seen, most structural model relationships do not differ between the two groups.

The only exceptions are the relationships between *COMP* and *CUSL*, *CUSA* and *CUSL*, as well as *LIKE* and *CUSL*, which differ significantly on a 10% level. More precisely, the effect between *COMP* and *CUSL* is significantly ($p \leq 0.10$) different between customers with a contract plan ($p^{(1)} = 0.136$) and those with a prepaid plan ($p^{(2)} = -0.062$). Similarly, the relationship between *CUSA* and *CUSL* is significantly ($p \leq 0.10$) different among customers with a contract plan ($p^{(1)} = 0.599$) versus those with a prepaid plan ($p^{(2)} = 0.440$). Finally, the effect of *LIKE* on *CUSL* is significantly ($p < 0.10$) different among customers with a contract plan ($p^{(1)} = 0.205$) versus those with a prepaid plan ($p^{(2)} = 0.424$).

To further analyze group-specific effects, we can run another multigroup analysis approach on the data. To do so, go to **Calculate → Multi-Group Analysis (MGA),** which you can find at the top of the SmartPLS screen. Alternatively, you can left-click on the wheel symbol in the tool bar and click the **Multi-Group Analysis (MGA)** option.

In the dialogue box that follows (Exhibit 4.15), we again need to specify the groups to be compared. To do so tick the box next to *GROUP_servicetype(1.0)* under **Groups A.** Next, select *GROUP_servicetype(2.0)* under **Groups B.** In the other tabs, we can specify settings related to the PLS-SEM algorithm, the bootstrapping procedure, missing value treatment, and variable weighting. Use the standard settings for all the options as described in, for example, Hair et al. (2017).

Exhibit 4.15 **Multigroup Analysis Dialogue Box in SmartPLS**

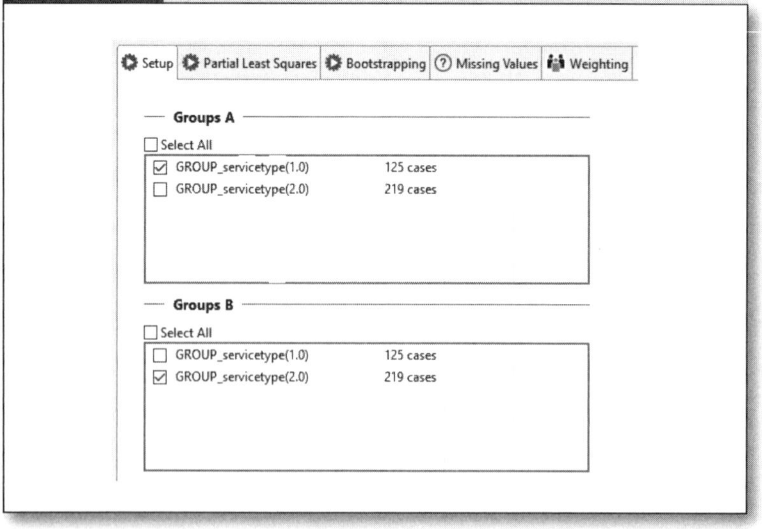

Exhibit 4.16	PLS-MGA Results in SmartPLS

Path Coefficients

Bootstrapping Results | Confidence Intervals (Bias Corrected) | PLS-MGA | Parametric Test | Welch-Satterthwait Test

	Path Coefficients-diff (| GROUP_servicetype(1.0) - GROUP_servicetype(2.0) |)	p-Value(GROUP_servicetype(1.0) vs GROUP_servicetype(2.0))
ATTR -> COMP	0.114	0.152
ATTR -> LIKE	0.112	0.209
COMP -> CUSA	0.106	0.228
COMP -> CUSL	0.197	0.039
CSOR -> COMP	0.118	0.865
CSOR -> LIKE	0.103	0.795
CUSA -> CUSL	0.159	0.027
LIKE -> CUSA	0.111	0.813
LIKE -> CUSL	0.220	0.977
PERF -> COMP	0.095	0.756
PERF -> LIKE	0.174	0.879
QUAL -> COMP	0.117	0.192
QUAL -> LIKE	0.165	0.120

In the results report that opens, go to **Final Results** → **Path Coefficients** and select the **PLS-MGA** tab to access the results of this multigroup analysis approach (Exhibit 4.16).

As PLS-MGA represents a one-tailed test, these p values indicate whether the path coefficient is significantly larger in the first group than in the second group. This is the case for the COMP → CUSL and CUSA → CUSL relationships. By taking 1-p value, we can also assess whether there is a significant difference in the other direction.

Exhibit 4.17	Parametric PLS Multigroup Tests Results

Path Coefficients

Bootstrapping Results | Confidence Intervals (Bias Corrected) | PLS-MGA | Parametric Test | Welch-Satterthwait Test

	Path Coefficients-diff (| GROUP_servicetype(1.0) - GROUP_servicetype(2.0) |)	t-Value...	p-Value...
ATTR -> COMP	0.114	1.027	0.305
ATTR -> LIKE	0.112	0.817	0.415
COMP -> CUSA	0.106	0.748	0.455
COMP -> CUSL	0.197	1.732	0.084
CSOR -> COMP	0.118	1.074	0.283
CSOR -> LIKE	0.103	0.860	0.390
CUSA -> CUSL	0.159	1.862	0.063
LIKE -> CUSA	0.111	0.878	0.381
LIKE -> CUSL	0.220	1.921	0.056
PERF -> COMP	0.095	0.721	0.472
PERF -> LIKE	0.174	1.193	0.234
QUAL -> COMP	0.117	0.847	0.397
QUAL -> LIKE	0.165	1.160	0.247

The results show that this is the case for the LIKE → CUSL relationship with a *p* value of 1 − 0.977 = 0.023. Note that the PLS-MGA approach relies on the bootstrapping procedure, which is a random process. Therefore, your test results will vary from those presented here, but the differences should not be substantial. Finally, Exhibit 4.17 presents the results of the parametric tests, which you can access by clicking the corresponding tab in the results report of the **Multi-Group Analysis (MGA)**. We show the results of the parametric test since the MICOM procedure revealed that the variances do not differ across groups. However, the Welch-Satterthwaite *t* test results, which you can find in the corresponding tab of the results report, are very similar.

In comparison, we find the PLS multigroup analysis results do not entail different outcomes across methods, as shown in Exhibit 4.18. Even though we recommend using the permutation test, the insights of such a multimethods approach provide additional confidence in the final results obtained.

Exhibit 4.18	PLS Multigroup Results Across Methods			
Path Coefficient	Permutation Test	PLS-MGA	Parametric Test	Welch-Satterthwaite t Test
ATTR → COMP				
ATTR → LIKE				
COMP → CUSA				
COMP → CUSL	X	X	X	X
CSOR → COMP				
CSOR → LIKE				
CUSA → CUSL	X	X	X	X
LIKE → CUSA				
LIKE → CUSL	X	X	X	X
PERF → COMP				
PERF → LIKE				
QUAL → COMP				
QUAL → LIKE				

Note: X indicates significant difference (*p* < 0.10) of path coefficients across the groups.

SUMMARY

- **Appreciate the importance of considering heterogeneity in PLS-SEM.** Applications of PLS-SEM are usually based on the assumption that the data represent a homogeneous population. In such cases, the estimation of a unique global PLS path model represents all observations. In many real-world applications, however, the assumption of sample homogeneity is unrealistic because respondents are likely to be heterogeneous in their perceptions and evaluations of latent phenomena. While the consideration of heterogeneity is promising from a practical and theoretical perspective to learn about differences between groups of respondents, it is oftentimes necessary, but also challenging, to obtain valid results. For example, when there are significant differences between path coefficients across groups, an analysis on the aggregate data level could cancel out group-specific effects. The results of such an analysis would likely be seriously misleading and render erroneous recommendations to managers and researchers. Researchers are well advised, therefore, to consider the issue of heterogeneous data structures in their modeling efforts.

- **Explain the difference between observed and unobserved heterogeneity.** Heterogeneity can be observed in that differences between two or more groups of data relate to observable characteristics, such as gender, age, or country of origin. Researchers can use these observable characteristics to partition the data into separate groups of observations and carry out group-specific PLS-SEM analyses. Second, heterogeneity can be unobserved in that it does not depend on one specific observable characteristic or combinations of several characteristics. To identify and treat unobserved heterogeneity, researchers can draw on a range of latent class techniques.

- **Comprehend the concept of measurement model invariance and its assessment in PLS-SEM.** Group comparisons are valid if measurement invariance has been established. Thereby, researchers ensure that group differences in model estimates do not result from the distinctive content and/or meanings of the latent variables across groups. When measurement invariance is not demonstrated, any conclusions about model relationships are questionable. The measurement invariances of composites (MICOM) procedure represents a useful tool in PLS-SEM. The procedure comprises three steps that test different aspects of measurement invariance: (1) configural invariance (i.e., equal parameterization and way of estimation), (2) compositional invariance (i.e., equal indicator weights), and (3) equality of composite mean values and variances.

- **Understand multigroup analysis and execute it in PLS-SEM.** Multigroup analysis allows testing whether differences between group-specific path coefficients are statistically significant. Researchers have proposed different approaches to multigroup analysis, with the *t* test–based parametric approach being most frequently applied. Alternatives such as the permutation test and the PLS-MGA do not rely on distributional assumptions. In comparison, the permutation test has particularly advantageous statistical properties, and we recommend this method when analyzing the difference of parameters across two groups. In the case of more than two groups, the use of the OTG is appropriate.

REVIEW QUESTIONS

1. What is the difference between observed and unobserved heterogeneity?

2. Explain the three steps of the MICOM procedure.

3. Why would you run a multigroup analysis in PLS-SEM?

4. Which approaches to multigroup analysis are available in PLS-SEM? Discuss their advantages and disadvantages.

CRITICAL THINKING QUESTIONS

1. Why is the consideration of heterogeneity so important when analyzing PLS path models?

2. Critically comment on the following statement: Measurement invariance is not an issue in PLS-SEM because of the method's focus on prediction and exploration.

3. What are the implications for your analysis when full measurement invariance holds?

4. Explain the conceptual differences between a multigroup analysis and an analysis of interaction effects.

KEY TERMS

Alpha inflation

Bonferroni correction

Bootstrap samples

Bootstrapping

Cluster analysis

Common factor model

Compositional invariance

Confidence interval

Configural invariance

Equality of composite
   mean values and
   variances

Familywise error rate

FIMIX-PLS

Finite mixture partial least
   squares (FIMIX-PLS)

Full measurement
   invariance

Latent class techniques

Measurement equivalence

Measurement invariance

Measurement invariance
   of composite models
   (MICOM) procedure

MICOM

Multigroup analysis

Multiple testing problem

Observed heterogeneity

Omnibus test of group
   differences (OTG)

OTG

Parametric test

Partial measurement
   invariance

Permutation

Permutation test

PLS-MGA

PLS-POS

Prediction-oriented
   segmentation in PLS-SEM
   (PLS-POS)

Šidák procedure

Unobserved heterogeneity

SUGGESTED READINGS

Becker, J.-M., Rai, A., Ringle, C. M., & Völckner, F. (2013). Discovering unobserved heterogeneity in structural equation models to avert validity threats. *MIS Quarterly*, *37*, 665–694.

Chin, W. W., & Dibbern, J. (2010). A permutation based procedure for multi-group PLS analysis: Results of tests of differences on simulated data and a cross cultural analysis of the sourcing of information system services between Germany and the USA. In V. Esposito Vinzi, W. W. Chin, J. Henseler, & H. Wang (Eds.), *Handbook of partial least squares: Concepts, methods and applications in marketing and related fields* (pp. 171–193). Berlin, Germany: Springer.

Diamantopoulos, A., & Papadopoulos, N. (2010). Assessing the cross-national invariance of formative measures: Guidelines for international business researchers. *Journal of International Business Studies, 61,* 1203–1218.

Hahn, C., Johnson, M. D., Herrmann, A., & Huber, F. (2002). Capturing customer heterogeneity using a finite mixture PLS approach. *Schmalenbach Business Review, 54,* 243–269.

Henseler, J., Ringle, C. M., & Sarstedt, M. (2016). Testing measurement invariance of composites using partial least squares. *International Marketing Review, 33,* 405–431.

Sarstedt, M., Henseler, J., & Ringle, C. M. (2011). Multi-group analysis in partial least squares (PLS) path modeling: Alternative methods and empirical results. *Advances in International Marketing, 22,* 195–218.

Schlägel, C., & Sarstedt, M. (2016). Assessing the measurement invariance of the four-dimensional cultural intelligence scale across countries: A composite model approach. *European Management Journal, 34,* 633–649.

C H A P T E R   5

Modeling Unobserved Heterogeneity

LEARNING OUTCOMES

1. Comprehend the FIMIX-PLS and PLS-POS approaches for identifying and treating unobserved heterogeneity.

2. Understand how to use FIMIX-PLS and PLS-POS jointly in a latent class analysis.

3. Appreciate how to execute a FIMIX-PLS analysis using SmartPLS.

4. Know how to use SmartPLS to run a PLS-POS analysis.

CHAPTER PREVIEW

As described in Chapter 4, researchers routinely use observable characteristics to partition data into groups and estimate separate models, thereby accounting for **observed heterogeneity**. However, the sources of heterogeneity in the data can hardly be fully known a priori. Consequently, situations arise in which differences related to **unobserved heterogeneity** prevent the derivation of accurate results as the analysis on the aggregate data level masks group-specific effects. Failure to consider heterogeneity can limit the validity of partial least squares structural equation modeling (PLS-SEM) results. Hence, identifying and—if necessary—treating unobserved heterogeneity is of crucial importance when using PLS-SEM. On the contrary, if unobserved heterogeneity does not substantially affect the results, researchers can analyze the data on the aggregate level, which is advantageous for the generalization of their outcomes.

In this chapter, we learn about two methods, finite mixture PLS (FIMIX-PLS) and prediction-oriented segmentation in PLS (PLS-POS), that enable researchers to identify and treat unobserved heterogeneity in PLS path models. Based on a description of the methods' basic working principles, we explain how they can be used jointly to identify the number of latent segments to retain from the data and estimate segment-specific path models. We also explain how to reproduce the latent segments by means of observable and managerially meaningful variables in order to turn the initial results into actionable understanding. The chapter concludes with a practical application of the methods drawing on our corporate reputation example and the SmartPLS software.

Uncovering Unobserved Heterogeneity in PLS Path Models

The usual way of treating heterogeneous data structures draws on standard **cluster analysis** methods such as **k-means clustering** (Sarstedt & Mooi, 2014). In the context of PLS-SEM, we could use k-means clustering on the indicator data or latent variables scores obtained from an initial analysis of the overall data set to form groups of data. The partition this analysis produces can then be used as input for group-specific PLS-SEM estimations. While easy to apply, a major drawback of such an approach is that these methods do not account for the measurement and structural model relationships as specified by the researcher. It is exactly these relationships, however, that are likely responsible for some of the group differences. At the same time, prior research has shown that traditional clustering approaches perform very poorly regarding identifying group differences (Sarstedt & Ringle, 2010). Acknowledging these limitations, methodological research in PLS-SEM has proposed a multitude of specific methods to identify and treat unobserved heterogeneity, commonly referred to as **latent class techniques** or **response-based segmentation techniques**. These techniques have proven very useful to identify unobserved heterogeneity and partition the data accordingly. Alternatively, latent class techniques may ascertain that unobserved heterogeneity does not influence the results, supporting an analysis of a single model based on the aggregate data level. Research has brought forward a multitude of different techniques that draw on, for example, genetic algorithm (Ringle, Sarstedt, & Schlittgen, 2014; Ringle, Sarstedt, Schlittgen, & Taylor, 2013), weighted least squares (Schlittgen, Ringle, Sarstedt, & Becker, 2015), and other approaches to identify unobserved heterogeneity using PLS-SEM (for a review, see Hair, Sarstedt, Matthews, & Ringle, 2016).

The first and most prominent latent class technique is **finite mixture partial least squares** (FIMIX-PLS; Hahn, Johnson, Herrmann, & Huber, 2002; Sarstedt, Becker, Ringle, & Schwaiger, 2011). As indicated by its name, the approach relies on the finite mixture models concept, which assumes that the overall population is a mixture of group-specific density functions. The aim of FIMIX-PLS is to disentangle the overall mixture distribution and estimate parameters (e.g., the path coefficients) of each group in a regression framework (i.e., mixture regressions; Becker, Ringle, Sarstedt, & Völckner, 2015; Wedel & Kamakura, 2000). Exhibit 5.1 shows an example of a mixture distribution that FIMIX-PLS aims to separate into segment-specific distributions.

Exhibit 5.1 | **Mixture Distribution Example**

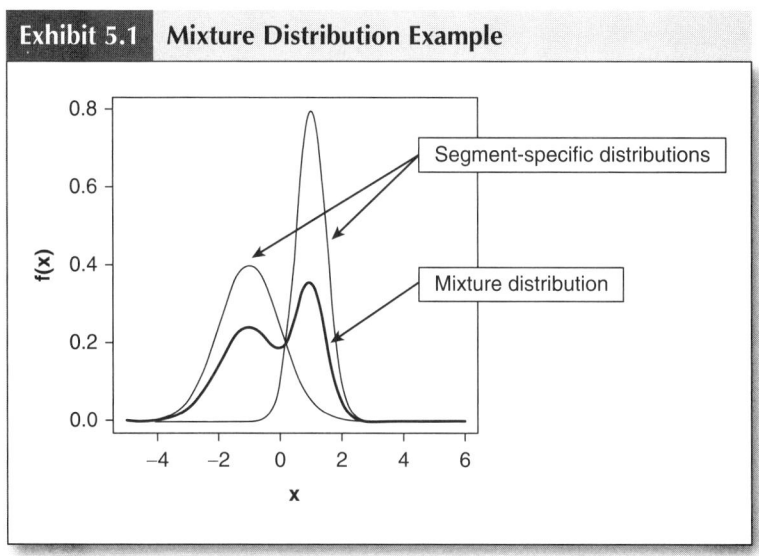

Source: Adapted from Hair, Sarstedt, Matthews, & Ringle (2016).

FIMIX-PLS follows two steps. In the first step, the standard PLS-SEM algorithm is run on the full set of data to obtain the scores of all the latent variables in the model. These latent variable scores then serve as input for a series of mixture regression analyses in the second step. The mixture regressions allow for the simultaneous probabilistic classification of observations into groups and the estimation of group-specific path coefficients. While the researcher needs to explicitly define the number of segments, FIMIX-PLS offers a range of statistical measures that provide an indication of how many segments likely underlie the data. This is a clear advantage of FIMIX-PLS as other PLS-SEM-based latent class techniques offer practically no insights in this respect. Simulation studies show that FIMIX-PLS reliably reveals the existence of heterogeneity in PLS path models and correctly indicates the appropriate number of segments to retain from the data (Sarstedt, Becker, et al., 2011). At the same time, however, FIMIX-PLS is clearly limited in terms of correctly identifying the underlying segment structure, as defined by the group-specific path coefficients (Ringle et al., 2013; Ringle et al., 2014), especially when the path model includes formative measures (Becker, Rai, Ringle, & Völckner, 2013). Furthermore, FIMIX-PLS is only capable of capturing heterogeneity in the structural model relationships and cannot account for heterogeneity in measurement models, which limits its usefulness for empirical research settings.

To overcome FIMIX-PLS's limitations, research has proposed a range of other alternatives. A very promising approach is Becker, Rai, Ringle, and Völckner's (2013) **prediction-oriented segmentation in PLS-SEM (PLS-POS)**, which follows a fundamentally different approach than FIMIX-PLS. Rather than defining heterogeneity at a distributional level (Exhibit 5.1), PLS-POS gradually reallocates observations from one segment to the other with the application of a goal criterion, which is the maximization of the explained variance provided by the segmentation solution. More specifically, PLS-POS computes each observation's distance to its own segment as well as other segments to decide on its group membership. When an observation has the shortest distance to its own segment, it remains in the current segment. Otherwise, the method (re-)assigns the observations to the alternative segment for which it exhibits the shortest distance. The algorithm repeatedly conducts this procedure for all the observations in the data set, thereby systematically improving the segment solution toward the goal criterion (i.e., maximization of the amount of explained variance across the segments). Becker et al.'s (2013) simulation study shows that PLS-POS reliably identifies segment structures and outperforms alternative segmentation techniques. However, PLS-POS does not offer concrete recommendations regarding the number of segments to consider in the analysis.

Against this background, and because of these approaches' joint implementation in SmartPLS 3, researchers benefit from using a combination of FIMIX-PLS and PLS-POS. The first step involves running FIMIX-PLS to identify the appropriate number of segments to retain from the data. The FIMIX-PLS solution then serves as a starting point for running PLS-POS thereafter. PLS-POS improves the FIMIX-PLS solution and also identifies heterogeneity in the measurement models. However, since the identified segments are—by definition—latent, the model estimates offer only limited guidance regarding the relationships to expect in the real world. The final step, therefore, involves reproducing the latent segments by means of observable and managerially meaningful variables in order to convert the initial results into actionable understanding. Exhibit 5.2 summarizes the steps in the systematic joint application of FIMIX-PLS and PLS-POS. Building on Hair et al. (2016) and Sarstedt, Ringle, and Gudergan (2016), in the following sections we discuss each step in greater detail.

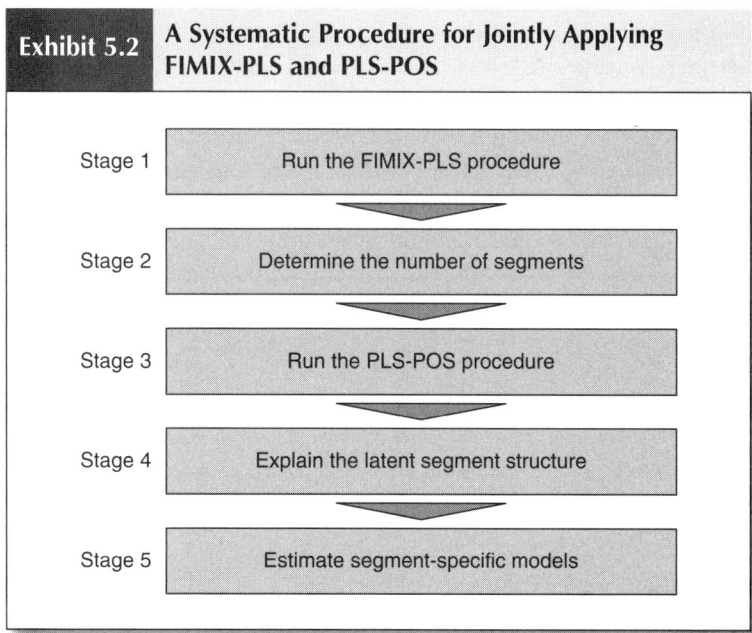

Exhibit 5.2 **A Systematic Procedure for Jointly Applying FIMIX-PLS and PLS-POS**

Stage 1   Run the FIMIX-PLS procedure

Stage 2   Determine the number of segments

Stage 3   Run the PLS-POS procedure

Stage 4   Explain the latent segment structure

Stage 5   Estimate segment-specific models

Step 1: Run the FIMIX-PLS Procedure

Running the FIMIX-PLS procedure requires the researcher to make several choices regarding the algorithm settings. The model estimation process in FIMIX-PLS follows the likelihood principle, which asserts that all of the evidence in a sample that is relevant for the model parameters is contained in the likelihood function. This likelihood function is maximized by using the **expectation-maximization (EM) algorithm.** The EM algorithm alternates between performing an expectation (E) step and a maximization (M) step. The E step creates a function for the expectation of the log-likelihood, which is evaluated using the current estimate of the parameters. The M step computes parameters by maximizing the expected log-likelihood identified in the E step. The E and M steps are successively applied until the results stabilize. Stabilization is reached when there is no substantial improvement in the (log) likelihood value from one iteration to the next. A threshold value of $1 \cdot 10^{-5}$ is recommended as a stop criterion to ensure that the algorithm converges at reasonably low levels of iterative changes in the log-likelihood values. When the stop criterion is set very low, the FIMIX-PLS algorithm may not converge within a reasonable time. Therefore, the researcher also needs to specify a maximum number of iterations after which the algorithm will automatically terminate. Specifying a maximum number of 5,000 iterations would normally ensure a sound balance between warranting acceptable computational running time and obtaining results that are precise enough.

Using the EM algorithm for model estimation is attractive because it is very efficient and always converges to a predefined number of segments. Convergence may occur in a local optimum, however, which means that the solution is only optimal compared to similar solutions but not globally (Steinley, 2003). To investigate the possible occurrence of a local optimum, researchers should run FIMIX-PLS multiple times. The initialization in FIMIX-PLS occurs randomly, which means that each time FIMIX-PLS is initiated the algorithm uses different randomly selected starting values for parameter estimation. Generally, the results of multiple FIMIX-PLS computations will be very similar. But if the results are not very similar, then a local optimum has occurred and the solution that is not consistent with the other solutions should be discarded. In line with simulation study results of the technique (Sarstedt, Becker, et al., 2011), we suggest using 10 repetitions of the FIMIX-PLS approach and choosing the solution with the best log-likelihood value. Another consequence of the algorithm's random nature is that the numbering of the segments is not determinate. That is, the results for one segment might appear in a different segment when FIMIX-PLS is run again. This characteristic is commonly referred to as **label switching** (McLachlan & Peel, 2000).

A further important consideration when running FIMIX-PLS involves the treatment of missing values. Kessel, Ringle, and Sarstedt (2010) have shown that just 5% missing values in one variable can cause severe problems in a FIMIX-PLS analysis when replacing them with the overall sample mean of that indicator's valid values (i.e., mean value replacement). In this case, the missing values treatment option creates a set of common scores, which FIMIX-PLS identifies as a distinct homogeneous segment. As a consequence, the number of segments will likely be overspecified and observations that truly belong to other segments will be forced into this artificially generated one. Therefore, mean value replacement must not be used in a FIMIX-PLS context, even if there are only very few missing values in the data set. Instead, researchers should remove all cases that include missing values in any of the indicators used in the model from the analysis (i.e., casewise deletion). While this approach also has its problems, particularly when values are missing at random (Sarstedt & Mooi, 2014), it avoids the generation of an artificial segment as is the case with mean value replacement and other imputation methods, such as EM imputation and regression imputation.

Finally, the FIMIX-PLS algorithm needs to be run for alternating numbers of segments, starting with the one-segment solution. As the number of segments is a priori unknown, researchers must compare the solutions with the different segment numbers in terms of their statistical adequacy and interpretability. The range of possible segment numbers depends on the interplay between the sample size and the minimum sample size requirements to reliably estimate the given model. For example, when analyzing a data set with 200 observations, and the requirement of a minimum segment sample size of 50, it is not reasonable to run FIMIX-PLS with more than four segments. It is therefore imperative to consider model-specific minimum segment sample size requirements (as documented, for example, in Hair, Hult, Ringle, & Sarstedt, 2017), before defining a range of segment solutions to consider in the FIMIX-PLS analysis. The theoretical maximum number of segments to consider is given by the largest integer, by dividing the sample size n by the minimum sample size. However, since it is highly unlikely that the observations are evenly distributed across the segments, especially when the upper bound is high, considering a lower number of segments is generally preferred.

Step 2: Determine the Number of Segments

A fundamental challenge with the application of FIMIX-PLS is determining the number of groups to retain from the data. Identifying a suitable number of groups is crucial, since many managerial decisions are based on this result. As Becker et al. (2015, p. 644) note, "a misspecified number of segments results in under- or oversegmentation, which can lead to inaccurate management decisions regarding, for example, customer targeting, product positioning, or determining the optimal marketing mix." FIMIX-PLS enables researchers to compute likelihood-based **information criteria** (also referred to as **model selection criteria**), which provide an indication of how many segments to retain from the data. Information criteria simultaneously take into account the fit (i.e., the likelihood) of a model and the number of parameters used to achieve that fit. The information criteria denote, therefore, a penalized likelihood function. That is, the negative likelihood plus a penalty term, which increases with the number of segments (Sarstedt, 2008). The smaller the value of a certain information criterion, the better the segmentation solution. Prominent examples of information criteria include **Akaike's Information Criterion** (AIC; Akaike, 1973), **Modified AIC with factor 3** (AIC$_3$; Bozdogan, 1994), **Consistent AIC** (CAIC; Bozdogan, 1987), and **Bayesian Information Criterion** (BIC; Schwarz, 1978). For a formal representation of these criteria see, for example, Sarstedt, Becker, et al. (2011).

Information criteria are not scaled within a certain range of values (e.g., between 0 and 1). Rather, the criteria may take values in the 100s or 1,000s, depending on the starting point of the FIMIX-PLS algorithm, which is set randomly. Importantly, however, each criterion's values can be compared across different solutions with varying segment numbers. Therefore, the researcher needs to examine several solutions with alternating numbers of segments and select the model that minimizes a particular information criterion.

Sarstedt, Becker, et al. (2011) have evaluated the efficacy of different information criteria in FIMIX-PLS across a broad range of data and model constellations. Their results demonstrate that researchers should jointly consider AIC_3 and CAIC. Whenever these two criteria indicate the same number of segments, this result likely indicates the appropriate number of segments. AIC with factor 4 (AIC_4; Bozdogan, 1994) and BIC generally perform well, while other criteria exhibit a pronounced overestimation tendency. This holds especially for AIC, which often overspecifies the correct number of segments by three or more segments. Still other criteria, such as Minimum Description Length 5 (MDL_5; Liang, Jaszak, & Coleman, 1992), show pronounced underestimation tendencies. Researchers can use this information to determine a certain range of reasonable segment numbers. For example, when AIC indicates a five-segment solution, retaining a smaller number of segments seems warranted. Exhibit 5.3 provides an overview of selected information criteria and highlights their performance in the context of FIMIX-PLS.

Information criteria are not a "silver bullet" to determine the most suitable number of segments in FIMIX-PLS, since criteria such as AIC_4 and BIC do not provide an indication of how well separated the segments are. For this reason, researchers should consider the complementary use of entropy-based measures, such as the **normed entropy statistic** (**EN**; Ramaswamy, DeSarbo, Reibstein, & Robinson, 1993). The EN uses the observations' segment membership probabilities to indicate whether the partition is reliable. The more observations exhibit high segment membership probabilities, the more clear-cut their segment affiliation will be. The EN ranges between 0 and 1, with higher values indicating a better quality partition. Prior research provides evidence that EN values above 0.50 permit a clear-cut classification of data into the predetermined number of segments (e.g., Ringle, Sarstedt, & Mooi, 2010).

When deciding on the number of segments to retain, it is particularly important to keep in mind that the EM algorithm always converges to the prespecified number of segments. The result may be, however, that FIMIX-PLS forces a small subset of data into extraneous segments, simply because the researcher selected a number of segments, which is too high. Such extraneous segments account for only a marginal portion of heterogeneity in the overall data set and are usually too small to ensure valid group-specific results (Rigdon, Ringle, & Sarstedt, 2010). Therefore, in addition to information criteria and the EN, the researcher should carefully consider the segment sizes produced by FIMIX-PLS. If the analysis yields an extraneous segment that is too small to warrant valid analysis, the researcher should consider reducing the number of segments or dropping this segment and focusing on the analysis and interpretation of the other, larger segments.

Exhibit 5.3	Selected Information Criteria and Their Performance in FIMIX-PLS	
Abbreviation	**Criterion Name**	**Performance in FIMIX-PLS**
AIC	Akaike's Information Criterion	• Weak performance • Very strong tendency to overestimate the number of segments • Can be used to determine the upper limit of reasonable segmentation solutions
AIC$_3$	Modified Akaike's Information Criterion With Factor 3	• Fair to good performance • Tends to overestimate the number of segments • Works well in combination with CAIC and BIC
AIC$_4$	Modified Akaike's Information Criterion With Factor 4	• Good performance • Tends to over- and underestimate the number of segments
BIC	Bayesian Information Criterion	• Good performance • Tends to underestimate the number of segments • Should be considered jointly with AIC$_3$
CAIC	Consistent Akaike's Information Criterion	• Good performance • Tends to underestimate the number of segments • Should be considered jointly with AIC$_3$
MDL$_5$	Minimum Description Length 5	• Weak performance • Very strong tendency to underestimate the number of segments • Can be used to determine the lower limit of reasonable segmentation solutions

Source: Adapted from Hair, Sarstedt, Matthews, & Ringle (2016, p. 70).

Finally, it is important to note that a purely data-driven approach provides only rough guidance regarding the number of segments to select. Heuristics such as information criteria and the EN have limitations, because they are sensitive to data and model characteristics. For example, research results by Becker et al. (2015) suggest that even low levels of collinearity in the structural model can have adverse consequences for the information criteria's performance. FIMIX-PLS is an exploratory tool and should be treated as such. Consequently, any decision regarding the number of segments should take practical considerations into account (e.g., Sarstedt, Schwaiger, & Ringle, 2009). For example, researchers might have a priori knowledge or a theory that can support their reasoning. Likewise, the number of segments must be small enough to ensure parsimony and manageability, but each segment should also be large enough to warrant strategic attention (Sarstedt & Mooi, 2014).

Step 3: Run the PLS-POS Procedure

Running the PLS-POS procedure requires the researcher to make several choices regarding the algorithm settings. In the following, we briefly explain the main steps of a PLS-POS analysis—as shown in Exhibit 5.4—and address the different choices to make. For a detailed explication of the PLS-POS algorithm, see Becker, Rai, Ringle, and Völckner (2013).

When starting the PLS-POS algorithm, the first choice to make is the number of segments to retain from the data. As indicated earlier, researchers should draw on the FIMIX-PLS results along with practical considerations to make this decision. Using the segment number as input, PLS-POS then randomly assigns each observation to one of the segments and continues with the analysis. Alternatively, PLS-POS can use the final partition resulting from the FIMIX-PLS analysis as input. That is, each observation is assigned to the segment where it has the highest probability of membership as indicated by FIMIX-PLS. This partition is then used as PLS-POS's starting partition.

In PLS-POS's initialization stage, researchers have the option of running a **presegmentation**. If this is chosen, PLS-POS uses the starting partition and then reassigns all observations at the same time to their closest segment in the first round—before reassigning observations one by one when running the subsequent iterations of the PLS-POS algorithm (Steps 2 and 3). Reassigning all observations at the

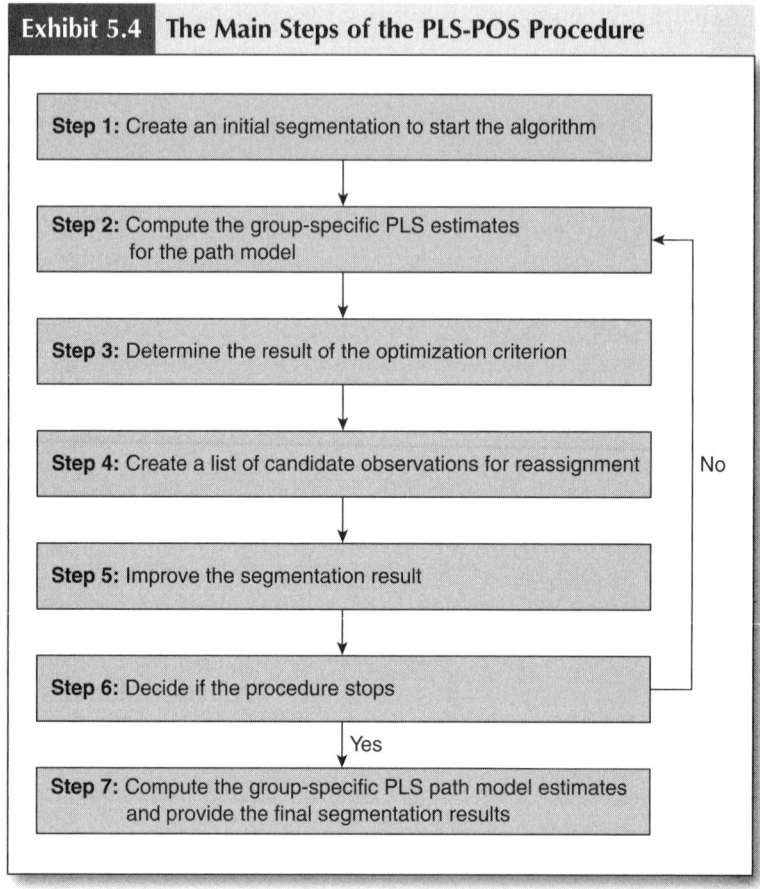

Exhibit 5.4 The Main Steps of the PLS-POS Procedure

Step 1: Create an initial segmentation to start the algorithm

Step 2: Compute the group-specific PLS estimates for the path model

Step 3: Determine the result of the optimization criterion

Step 4: Create a list of candidate observations for reassignment

No

Step 5: Improve the segmentation result

Step 6: Decide if the procedure stops

Yes

Step 7: Compute the group-specific PLS path model estimates and provide the final segmentation results

same time in the first round does not guarantee that all changes contribute to improving the optimization criterion. However, this option may save some time since it usually entails a significant improvement of the starting partition. The presegmentation proves valuable only when PLS-POS initializes the starting partition randomly. In this case, researchers should run PLS-POS several times since the quality of the final solution depends on the starting partition. Similarly to FIMIX-PLS, select the best solution—regarding the final value of the optimization criterion—of 10 repetitions to avoid convergence on a local optimum. However, when using the FIMIX-PLS solution as the starting partition, we recommend not using the presegmentation option as this step entails considerable changes in the FIMIX-PLS solution. As the FIMIX-PLS solution already converged in the global or a local optimum solution, the changes that come with the presegmentation would distort the FIMIX-PLS analysis.

Also, researchers need to decide on the **search depth**, which determines the maximum number of candidate observations for segment reassignment (see Step 4). If the search depth has a value of, for example, 10, PLS-POS analyzes only the first 10 candidates in the sorted list of observations for reassignment. Usually, researchers should analyze all candidate observations and, therefore, use a search depth equal to the number of observations. Nevertheless, reducing the search depth may be advantageous in complex segmentation settings (i.e., a large number of observations and/or a high number of segments) to reduce computation time, which can become excessive. However, for the final PLS-POS computations, we recommend using a search depth equal to the number of observations in the full data set.

Another decision to make in the initialization of the PLS-POS algorithm concerns the choice of the **optimization criterion**. Consistent with the PLS-SEM algorithm's predictive nature (Lohmöller, 1989; Wold, 1982, 1985), the objective of PLS-POS is to form groups of observations that maximize the prediction of the endogenous latent variables in their path model estimations. For this purpose, a suitable optimization criterion is to maximize the sum of the endogenous latent variables' explained variance (i.e., their R^2 values) across all the groups (clustering for maximum prediction; Anderberg, 1973). Alternatively, researchers can use the weighted sum of each group's sum of R^2 values as the optimization criterion, whereby the relative segment sizes of the different groups serve as the weights. As a consequence, the R^2 values of the larger segments contribute more strongly to the computation of the optimization criterion compared to smaller segments. Finally, researchers can focus the optimization criterion only on a particularly important endogenous latent variable in the PLS path model. Examples of such key target constructs are customer satisfaction and customer loyalty in customer satisfaction models (e.g., Fornell, Johnson, Anderson, Cha, & Bryant, 1996) and intention to use in technology acceptance models (e.g., Al-Ghantani, Hubona, & Wang, 2007). Nevertheless, we generally recommend considering all endogenous latent variables, using the weighted sum of each group's sum of R^2 values as the optimization criterion.

In Steps 2 and 3, PLS-POS estimates the group-specific PLS path models and uses the results as input for the computation of the optimization criterion. In Step 4, the method identifies a set of candidate observations for reassignment from their current segment to an alternative segment in order to improve the optimization criterion. Following Squillacciotti (2005, 2010), PLS-POS employs a distance measure to reassign observations: The closer the distance of an observation to a segment, the higher the predictive ability of this observation with respect to that segment. The distance measure draws on the structural model residuals, which are the differences between the predicted and actual scores of the endogenous latent variable(s). More precisely, the distance measure uses the observations' **prediction errors**, which are equal to the squared structural model residuals. Note that the residuals are squared so that larger differences between predicted and actual scores contribute more strongly to the prediction error than small differences. The sum of squared residuals (i.e., the prediction errors) represents the distance value.

While the actual endogenous latent variable scores result from the group-specific PLS-SEM analyses, the predicted scores need to be computed separately. This is done by multiplying the observations' exogenous latent variable scores with the corresponding path coefficient estimates and computing the sum of these products. To illustrate the computation of the predicted latent variable scores and the prediction error, consider the example of two equally sized segments and their segment-specific path coefficient estimates as shown in Exhibit 5.5. Furthermore, consider an observation belonging to segment 1 with (standardized) latent variable scores of 0.1 for construct Y_1, 0.6 for construct Y_2, and 0.5 for construct Y_3. To determine how well this observation predicts the target construct Y_3 in its own segment, PLS-POS multiplies the observation's latent variable score for Y_1 with the path coefficient p_1 (i.e., $0.1 \cdot 0.7 = 0.07$) as well as the observation's latent variable score for Y_2 with the path coefficient p_2 (i.e., $0.6 \cdot 0.2 = 0.12$). Adding these two products yields the observation's predicted latent variable for construct Y_3 of $0.07 + 0.12 = 0.19$. The actual latent variable score of Y_3 for this observation is 0.5, yielding a prediction error of $(0.19 - 0.5)^2 = 0.0961$. Correspondingly, this observation's prediction error for segment 2 is given by $[(0.1 \cdot 0.2 + 0.6 \cdot 0.7) - 0.5]^2 = 0.0036$. We find that the observation has a higher prediction error in its own segment 1, compared to segment 2. In other words, the observation has a higher predictive ability in segment 2 because the prediction error is smaller compared to segment 1. The sum of squared residuals (or prediction

Exhibit 5.5	Example of the Prediction Error Computation

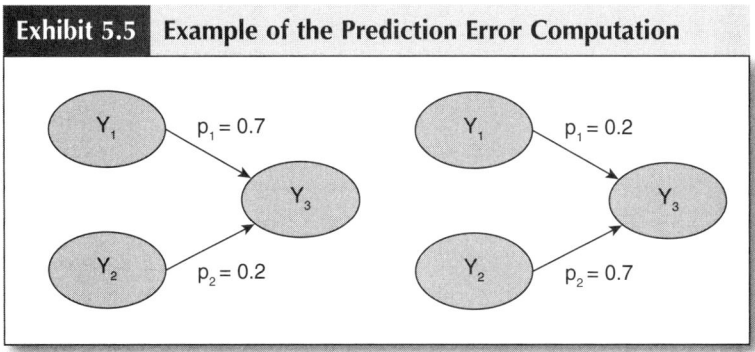

errors) represents each observation's final distance values to its current segment and possible alternative segment assignments.

Based on the observations' distance values, PLS-POS generates a list of candidate observations for reassignment in Step 4 of the algorithm (Exhibit 5.4). If an observation has the shortest distance to its own segment, this observation does not become a candidate for reassignment. On the contrary, if an observation is closer to another than its own segment (as explained for the example shown in Exhibit 5.5), it fits better into that segment in terms of predictive ability, thereby potentially improving the optimization criterion after reassignment. If an observation has a shorter distance to more than one other segment, the segment with the shortest distance is considered in the list. Subsequently, the candidate observations are sorted in descending order of distances. If the list of candidates for reassignment is empty, the algorithm proceeds to Step 7 and stops with the final results computation (Exhibit 5.4).

In Step 5, PLS-POS reassigns the first observation in the list of candidates into the alternative segment to which it has the shortest distance. A key challenge of this approach is the indeterminacy of the data assignment task as it is unknown how the group-specific PLS results will change after an observation is reassigned to a different group. For this reason, the PLS-POS algorithm exchanges only one observation at a time. For the changed data groups, the method updates the segment-specific PLS-SEM results and the optimization criterion. If the optimization criterion improves, the algorithm maintains the reassignment and proceeds to Step 6. Otherwise, if the optimization criterion does not improve, the algorithm does not reassign the observation. Then, the same assessment continues with the next candidate observation. As a result, the PLS-POS algorithm reassigns an observation only if this step improves the new segmentation solution's optimization criterion.

In Step 6, PLS-POS checks whether the maximum number of iterations has been reached. If not, the algorithm continues with the next iteration starting with Step 2. Otherwise, the algorithm continues with Step 7, the final step, which computes the PLS-POS results of the segmentation solution including the group-specific PLS path model estimates. The maximum number of iterations should be sufficiently high to obtain a solution that is close to the global optimum. We suggest using a maximum number of iterations equal to the higher of the following: 1,000 or two times the number of observations.

In summary, PLS-POS accounts for heterogeneity in the PLS path model by using a distance measure that facilitates the reassignment of observations in an effort to improve the prediction-oriented optimization criterion. The method is generally applicable to all PLS path models regardless of the mode of measurement model (for more details, see Becker, Rai, Ringle, & Völckner, 2013), the distribution of the data, or the complexity of the structural model. Like the EM algorithm in FIMIX-PLS, PLS-POS faces the problem that it may result in a local optimum solution due to its use of a hill-climbing approach. Thus, if one does not build on the FIMIX-PLS starting partition, a repeated application of PLS-POS (e.g., 10 times per prespecified number of segments) with different starting partitions is advisable.

Step 4: Explain the Latent Segment Structure

Upon completion of the analysis, PLS-POS provides users with each observation's group membership along with the segment-specific model estimates for the measurement and structural models. As the identified segments are—by definition—latent, the model estimates do not explain what makes up these latent segments. Turning the initial results into actionable understanding requires the researcher to interpret the segments in terms of observable and managerially meaningful variables. To do so, researchers need to translate the latent segment structure into an observable one by partitioning the data using one or more explanatory variable(s). The resulting partition of data should largely correspond to the grouping produced by PLS-POS. Such an analysis is also referred to as **ex post analysis**.

An ex post analysis may involve running a binary (in case of two latent segments) or multinomial (in case of three or more latent segments) logistic regression with the segment affiliation as the dependent variable and potential explanatory variables as the independent variables (e.g., Money, Hillenbrand, Henseler, & Da Camara, 2012; Wilden & Gudergan, 2015). Comparing the resulting log odds allows identifying those independent variables that have the strongest effect on the latent segment structure. Using cross tabs, one can then contrast the frequency distribution of the PLS-POS segment affiliation variable with one or more explanatory variables. Exhibit 5.6 shows a cross tab example for 400 observations, two PLS-POS segments, and the explanatory variable *gender*. The high frequencies indicate that the vast majority of respondents in PLS-POS segment 1 are female. On the contrary, PLS-POS segment 2 primarily includes male respondents. More precisely, the variable *gender* explains 188 + 177 = 365 of 400 PLS-POS segment assignments, which is an overlap of 365 / 400 = 91.25%. On the contrary, if each cell included 100 respondents, gender would be unsuitable for explaining the PLS-POS segmentation. In general, the more equal the distribution across the cells, the less useful is the explanatory variable and vice versa.

Researchers have also drawn on tree-building algorithms such as Chi-squared Automatic Interaction Detector (CHAID) or classification and regression trees (CART) to identify suitable explanatory variables (e.g., Ringle et al., 2010; Sarstedt & Ringle, 2010). These techniques can be used to predict values of a categorical variable (in our case the observations' latent segment affiliations) from one or more continuous and/or categorical predictor variables. Both CHAID and CART techniques will construct trees, where each (nonterminal) node identifies a split condition, to yield an optimum classification. The resulting nodes can then be used to identify combinations of explanatory variables that exhibit a large overlap with the latent class segmentation.

Importantly, to successfully run an ex post analysis, researchers must be able to consider a wide range of observable characteristics that can serve as possible input. Examining too few observable characteristics restricts the researcher's ability to reproduce the

Exhibit 5.6	Cross Tab Example		
	Female	*Male*	*Sum*
PLS-POS segment 1	188 (47.00%)	14 (3.50%)	**202 (50.50%)**
PLS-POS segment 2	21 (5.25%)	177 (44.25%)	**198 (49.50%)**
Sum	**209 (52.25%)**	**191 (47.75%)**	**400 (100.00%)**

PLS-POS partition. With this in mind, researchers should assess whether a single explanatory variable, or a set of variables, has theoretical meaning regarding elucidating possible differences in path coefficients across identified segments. Therefore, assessing the explanatory role of possible variables, so that the latent class analysis can be implemented more completely, should be considered in the research design stage when collecting descriptive or other variables that may matter.

Nevertheless, reproducing the PLS-POS partition remains a very challenging task, as observable characteristics often do not match the latent segment structures well. Against this background, an overlap of 60% between the PLS-POS partition and the one produced by the explanatory variable(s) can be considered satisfactory.

Step 5: Estimate Segment-Specific Models

Once the researcher has identified one or more explanatory variables that match the PLS-POS partition well, the final step is to estimate segment-specific models as indicated by the explanatory variable(s). In doing so, the researcher must ensure that all the model measures meet common quality standards as documented in, for example, Hair et al. (2017). These analyses complete the latent class analysis. However, further analyses may involve testing whether the numerical differences between segment-specific path coefficients are also significantly different by means of multigroup analysis, which should also include testing for measurement invariance as described in Chapter 4. Exhibit 5.7 summarizes the rules of thumb for a latent class analysis in PLS-SEM.

CASE STUDY ILLUSTRATION—LATENT CLASS ANALYSIS

Step 1: Run the FIMIX-PLS Procedure

Our illustration of the latent class analysis is taken partially from Matthews, Sarstedt, Hair, and Ringle's (2016) case study on the use of FIMIX-PLS, which also draws on the extended corporate reputation model. In light of the problems that may arise when using mean imputation to handle missing data, we deleted all observations with missing values in the data examination stage (i.e., casewise deletion) prior to the actual analysis. After casewise deletion, 336 observations remain.

Exhibit 5.7	**Rules of Thumb for a Latent Class Analysis**

FIMIX-PLS algorithm settings:

- Use a stop criterion of $1 \cdot 10^{-5}$ and a maximum number of 5,000 iterations.

- Use the best solution of 10 repetitions to avoid convergence on a local optimum.

- When the indicators have missing values, use casewise deletion. Do not use mean value replacement or any other imputation method.

- To define a range of reasonable segment numbers, use one segment as the lower bound and the largest integer when dividing the sample size by the minimum segment sample size as the upper bound.

Determining the number of segments to retain:

- Information criteria: If AIC_3 and $CAIC$ indicate the same number of segments, choose this solution. Alternatively, jointly consider AIC_3 and BIC. Also consider the segment number as indicated by AIC_4 and BIC. Generally, choose fewer segments than indicated by AIC and more segments than indicated by MDL_5.

- The EN should preferably be higher than 0.5.

- Ensure that the segment sizes meet the minimum sample size requirements. If not met, reduce the number of segments, or discard the extraneous segments and focus on the remaining larger ones.

- Take practical considerations into account. If possible, let a priori information and theory contribute to substantiating segments. Ensure that the solution is managerially relevant.

PLS-POS algorithm settings:

- Use the FIMIX-PLS results to determine the number of segments to retain from the data. Consider taking the FIMIX-PLS solution as the starting partition for PLS-POS.

- Do not use the presegmentation option when using the FIMIX-PLS solution as the starting partition for PLS-POS.

- When using random assignment as the starting partition, use the best solution of 10 repetitions to avoid convergence in a local optimum.

- Select a high maximum number of iterations that the PLS-POS algorithm will perform. This number should equal the higher of the following: 1,000 or two times the number of observations.

- The search depth should equal the number of observations.

- Consider using the weighted sum of all constructs' R^2 values as the optimization criterion.

(Continued)

Exhibit 5.7 (Continued)

Ex post analysis:

- Partition the data by using an explanatory variable, or a combination of several explanatory variables, which yields a grouping of data that largely corresponds to the one produced by PLS-POS.
- A 60% overlap between the PLS-POS partition and the one produced by the explanatory variable(s) is considered satisfactory. Consider running a PLS multigroup analysis after establishing (partial) measurement invariance.

This reduced data set is already included in the SmartPLS Project with the name **Corporate Reputation – Advanced PLS-SEM Book,** which we use for all case study illustrations in this book (see Chapter 1) and is available at www.pls-sem.com. Navigate to the SmartPLS **Projects** window on the left-hand side. Right-click on the data set **Corporate reputation data casewise deletion [336 records]** and choose **Select Active Data File** from the menu (Exhibit 5.8). Next, double-click on **Extended model,** and the extended PLS path model for the corporate reputation example opens in the SmartPLS modeling window. Note that you obtain the same results when using the full data set with 344 observations and selecting the **Casewise Deletion** in the **Missing Values** tab in the starting dialogue when running the segmentation algorithms. However, for better illustration purposes, we use the **Corporate reputation data reduced** with 336 observations.

To initiate the FIMIX-PLS analysis, click on **Calculate → Finite Mixture (FIMIX) Segmentation** in the menu bar. Alternatively, you can left-click on the wheel symbol in the tool bar and select the corresponding option in the combo box that opens. After selecting the FIMIX-PLS option, the dialogue box in Exhibit 5.9 appears. For the initial analysis, start with a one-segment solution and use the default settings for the stop criterion ($1 \cdot 10^{-5}$), maximum number of iterations (**5,000**), and the number of repetitions (**10**). The dialogue box has two further tabs to specify the standard PLS-SEM algorithm settings and the treatment of missing values. Use the default setting for the PLS-SEM algorithm. Finally, click on **Start Calculation.**

After convergence, SmartPLS 3 opens a new report tab, which shows the results of the FIMIX-PLS analysis. Before analyzing the results in detail, we need to rerun FIMIX-PLS for higher-segment

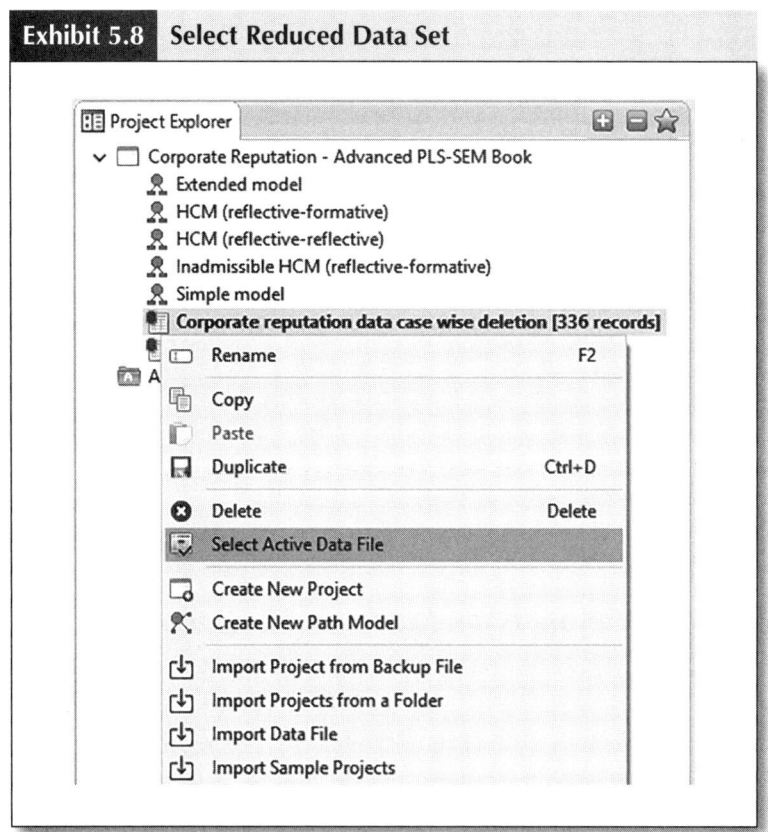

Exhibit 5.8   Select Reduced Data Set

solutions (e.g., two to five segments). To determine the upper bound of the range of segment solutions to consider, check the minimum sample size requirements as specified in Hair et al. (2017). With a maximum number of eight arrowheads pointing at any construct in the model (formative indicators of $QUAL$) and assuming a 5% significance level, as well as a minimum R^2 of 0.25, we would need 54 observations to reliably estimate the model. The greatest integer from dividing the sample size (i.e., 336) by the minimum sample size (i.e., 54) yields a theoretical upper bound of 6.22 = 6. However, given the complexity of the model, we consider only a one- to five-segment solution, especially since an equal distribution of observations— which would be necessary to meet the minimum sample size requirements—is highly unlikely. Thus, we rerun FIMIX-PLS for two to five segments, using the above algorithm settings, and save the results report for each run.

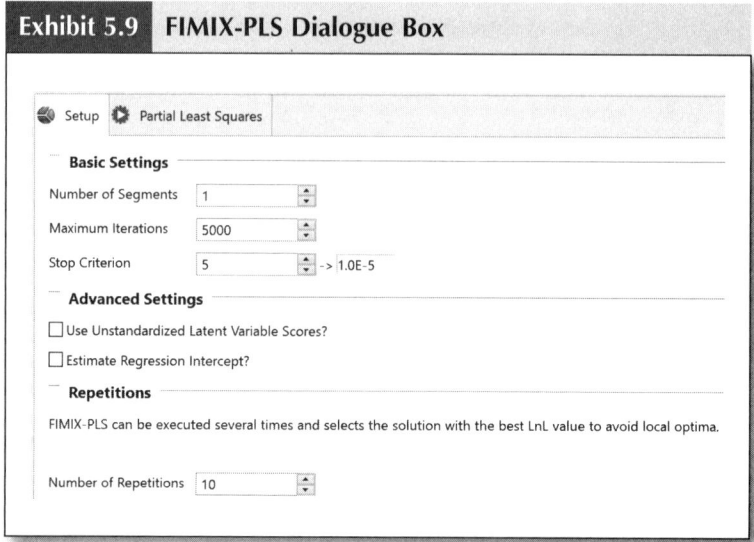

Exhibit 5.9 FIMIX-PLS Dialogue Box

Step 2: Determine the Number of Segments

To determine the number of segments to retain from the data, we need to examine the fit indices, which you can find under **Quality Criteria → Fit Indices** in each of the FIMIX-PLS results reports. To facilitate their comparison across the different segment number solutions, you can export the values to a spreadsheet program such as Microsoft Excel. To do so, click on **Export to clipboard: CSV** in the **Fit Indices** tab, and paste the values into an Excel file. Exhibit 5.10 provides an overview of the log-likelihood (LnL) values, information criteria, and the EN for a one- to five-segment solution.

Two aspects are worth remembering: First, your values will look different from those reported in Exhibit 5.10, because FIMIX-PLS initializes randomly and therefore produces different results each time it is run. Nevertheless, the conclusions should not differ fundamentally regarding the number of segments to retain. Second, remember that for each fit measure, the optimal solution is the number of segments with the lowest value (see bold numbers in Exhibit 5.10), except in terms of EN, where higher values indicate a better separation of the segments.

Unfortunately, AIC_3 and CAIC do not indicate the same number of segments, and neither do AIC_3 and BIC. AIC shows a five-segment

Exhibit 5.10	Fit Indices for a One- to Five-Segment Solution				
	Number of Segments				
Criteria	*1*	*2*	*3*	*4*	*5*
LnL	−1,406.86	−1,353.83	−1,319.79	−1,290.62	−1,261.28
AIC	2,847.72	2,777.66	2,745.58	2,723.2	**2,700.57**
AIC_3	2,864.72	2,812.66	2,798.58	2,794.24	**2,789.57**
AIC_4	2,881.72	**2,847.66**	2,851.58	2,865.24	2,878.57
BIC	2,912.61	**2,911.26**	2,947.89	2,994.25	3,040.29
CAIC	**2,929.61**	2,946.26	3,000.89	3,065.25	3,129.29
MDL_5	**3,308.18**	3,725.65	4,181.11	4,646.31	5,111.18
EN	n/a	0.43	0.64	0.67	**0.70**

solution, suggesting that the correct number likely is lower than this. On the other extreme, CAIC and particularly MDL_5 both result in a one-segment solution, suggesting that two or more segments should be considered. Further analysis reveals that two other criteria with good performance in detecting an appropriate number of segments, AIC_4 and BIC, both indicate two segments, thus providing initial support for this solution. However, the two-segment solution exhibits an EN value below 0.50, suggesting that the two segments are not well separated.

In the results report, under **Final Results → Segment Sizes**, SmartPLS 3 shows the relative segment sizes of the FIMIX-PLS solution for a certain prespecified number of segments. Examining the relative segment sizes across FIMIX-PLS solutions for different numbers of segments, as summarized in Exhibit 5.11, shows that selecting more than two segments is not reasonable. Note that these segment sizes do not result from a hard clustering of observations based on the maximum probabilities of segment membership but from weighted least squares regressions using these probabilities as input. For example, for a three-segment solution, the breakdown of segment sizes is segment 1: 52% (52% of 336 = 175), segment 2: 41.2% (41.2% of 336 = 138), and segment 3: only 6.8% (6.8% of 336 = 23). As can be seen, with 23 observations, segment 3 is too small for a segment-specific PLS-SEM analysis. One way to handle this is to consider dropping the third segment as it is too small to warrant valid analysis, and instead focus on the analysis and interpretation of the two other, larger segments.

Exhibit 5.11	Relative Segment Sizes				
Number of Segments	*Relative Segment Sizes*				
	Segment 1	*Segment 2*	*Segment 3*	*Segment 4*	*Segment 5*
2	0.515	0.485			
3	0.520	0.412	0.068		
4	0.526	0.354	0.085	0.035	
5	0.477	0.282	0.106	0.093	0.043

The SmartPLS 3 results report returns the R^2 values of the FIMIX-PLS solution under **Quality Criteria → R Square.** Exhibit 5.12 shows the segment-specific R^2 values along with the weighted R^2 values that result from multiplying the segment-specific values with the relative segment sizes. The R^2 values in segment 1 are considerably higher compared to the full set of data, while those in segment 2 are smaller. We also find that the weighted average R^2 values of the FIMIX-PLS two-segment solution are only slightly higher than those of the full set of data. These results substantiate our concerns regarding the FIMIX-PLS segmentation solution: We may have a sample that potentially is homogenous. On the other hand, when inspecting the FIMIX-PLS segment-specific path coefficients in the results report under **Final Results → Path Coefficients,** we find that some differences may be relevant (i.e., in their size) and interesting for forming and interpreting different groups of data. For example, the strengths of the standardized path coefficient from *CUSA* to *CUSL* differs by more than 0.15 points in the two segments solution.

Overall, in combination these results suggest that heterogeneity may not be prevalent in our study. The fit indices yield ambiguous results regarding the number of segments to retain from the data while the relative segment sizes permit only a two-segment solution. The examination of the segment-specific R^2 values supports this conclusion as the weighted average R^2 values are only slightly higher than the R^2 values produced when drawing on the full set of data. Only the segment-specific path coefficients promise some interesting but relatively small differences. Thus, using different segments for further analyses is likely not beneficial. However, for illustrative purposes to

Exhibit 5.12	FIMIX-PLS R^2 Values for the Two-Segment Solution			
	R^2 values			FIMIX-PLS
	Full Data Set	FIMIX-PLS Segment 1	FIMIX-PLS Segment 2	Weighted Average R^2 Values
COMP	0.629	0.812	0.519	0.670
CUSA	0.293	0.530	0.164	0.352
CUSL	0.562	0.807	0.389	0.604
LIKE	0.557	0.814	0.361	0.594

outline how PLS-POS can be applied, we take a more detailed look at the two-segment solution using PLS-POS.

Step 3: Run the PLS-POS Procedure

In the next step, we run the PLS-POS procedure by clicking on **Calculate** → **Prediction-Oriented Segmentation** (POS) in the menu bar. Alternatively, you can left-click on the wheel symbol in the tool bar and select the corresponding option in the combo box that opens. Then, the dialogue box in Exhibit 5.13 appears. For the number of **Groups**, select 2—as revealed by the previous FIMIX-PLS analysis. Since the data set has 336 observations after casewise deletion and 2 · 336 = 672, which is lower than 1,000, select the default value of **1,000** for the **Maximum Number of Iterations**. The **Search Depth** should equal the number of observations, which is 336. If the default value of **1,000** remains unchanged, SmartPLS automatically stops when 336 has been reached since it is not possible to search beyond the maximum number of observations in the data set. In the **Advanced Settings**, we select the **FIMIX-PLS Segmentation** instead of the **Random Assignment** option for the **Initial Separation** in PLS-POS and do not opt for applying the **Pre-Segmentation** option. Furthermore, we do not want to focus on a single endogenous latent variable in the corporate reputation model such as *CUSL*. Instead, we want to maximize the group-specific explanation of all endogenous latent variables at the same time. We therefore select the **Sum of all Construct Weighted R-Squares** as the **Optimization Criterion**. Finally, click on **Start Calculation**.

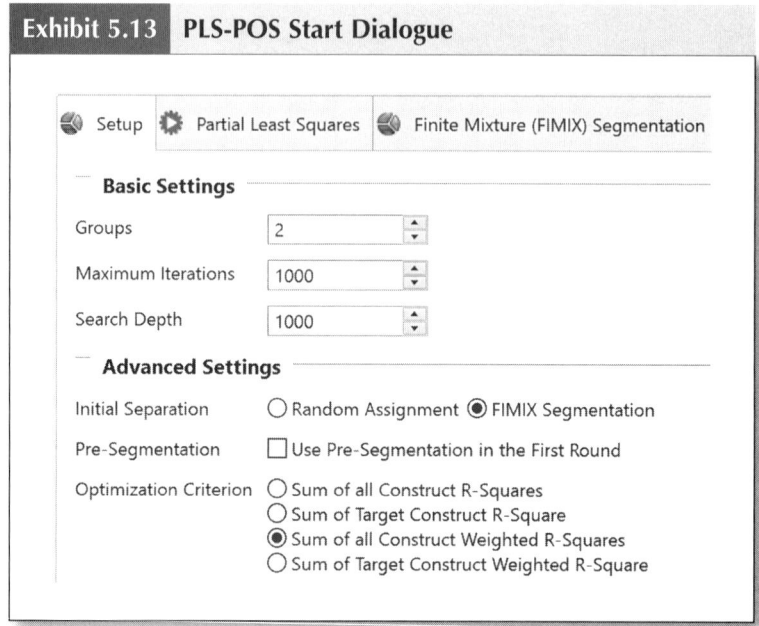

Exhibit 5.13 **PLS-POS Start Dialogue**

After convergence, SmartPLS 3 opens a new **Report** tab, which shows the results of the PLS-POS analysis. Before analyzing the results in detail, we need to rerun PLS-POS to obtain 10 solutions. In the PLS-POS results **Report**, go to **Quality Criteria → Change in Objective Criterion** and select the solution with the best solution (i.e., highest objective value outcome). For this solution, we begin the results analysis with the R^2 values—by navigating to **Quality Criteria → R Square**— and the (relative) segment size by going to **Final Results → Segment Sizes** in the **Report**. Exhibit 5.14 summarizes the PLS-POS results of the two-segment solution. While segment 1 (relative segment size: 56.8%) returns substantially higher R^2 values, the results for segment 2 (relative segment size: 43.2%) are at similar and higher levels compared to the outcomes for the full data set. Importantly, the average weighted R^2 values of the PLS-POS solution for two segments is considerably higher than the R^2 values of the full data set. This outcome is highly desirable for a segmentation solution.

Exhibit 5.15 shows the segment-specific path coefficients from the PLS-POS analysis. We find substantial differences in the path coefficient estimates across the two groups. For example, while *COMP* has a pronounced positive impact on *CUSL* in the first segment (0.242), the same effect is negative in the second segment (−0.302).

On the contrary, in the full data set, the corresponding path coefficient is close to zero (0.011).

Exhibit 5.14	PLS-POS Segmentation Solution			
	R^2 values			PLS-POS
	Full Data Set	PLS-POS Segment 1	PLS-POS Segment 2	Weighted Average R^2 Values
COMP	0.629	0.770	0.618	0.704
CUSA	0.293	0.521	0.421	0.478
CUSL	0.562	0.706	0.555	0.641
LIKE	0.557	0.631	0.657	0.642

Exhibit 5.15	Path Coefficients of the Full Set of Data and the Two PLS-POS Segments			
	Full Data Set	Segment 1	Segment 2	Segment 1– Segment 2*
ATTR → COMP	0.085	0.206	−0.115	0.321
ATTR → LIKE	0.159	0.144	0.113	0.031
COMP → CUSA	0.135	0.673	−0.438	1.112
COMP → CUSL	0.011	0.242	−0.302	0.544
CSOR → COMP	0.057	−0.129	0.194	−0.323
CSOR → LIKE	0.190	0.267	0.168	0.099
CUSA → CUSL	0.510	0.586	0.181	0.405
LIKE → CUSA	0.445	0.069	0.808	−0.739
LIKE → CUSL	0.334	0.096	0.747	−0.651
PERF → COMP	0.295	0.397	0.175	0.223
PERF → LIKE	0.116	0.267	−0.114	0.381
QUAL → COMP	0.431	0.455	0.588	−0.133
QUAL → LIKE	0.378	0.223	0.692	−0.470

* Instead of the frequently used absolute difference, we use the segment 1 – segment 2 difference of the segments' specific path coefficients. As a result, a positive sign of the difference indicates that the path coefficient of segment 1 is larger than that of segment 2 and vice versa.

Step 4: Explain the Latent Segment Structure

In order to be able to interpret the segments in terms of observable and practically meaningful variables, we next run an ex post analysis to identify one or more explanatory variable(s) that match the PLS-POS partition in the best possible way. To begin with, we first have to transfer the partition as indicated by PLS-POS to the original data set. To do so, open the original data set with 336 observations in Microsoft Excel or some other spreadsheet software. Next, go back to the SmartPLS **Results** report of PLS-POS and display **Final Results** → **Group Assignment**. Use the **Copy to Clipboard** → **Excel Format** option in the upper right-hand corner of the results view and paste the group assignment to previously used Microsoft Excel sheet (i.e., to the next empty column on the right-hand side of your data). Delete the column with the "unit" information and rename the variable "group" into "PLS-POS Groups." The results should look like Exhibit 5.16. Save the data set with an intuitive name (e.g., "Corporate reputation data 336 pls-pos") in the comma separated value (csv) data format.

We can now use this data set to compare the PLS-POS partition with those indicated by other observable variables in the data set. Unfortunately, the corporate reputation data set has only two such variables, which indicate each respondent's service provider (service provider 1–4) and the type of service the respondent uses (prepaid or contract). Therefore, the chances of reproducing the PLS-POS partition adequately are relatively low. We strongly recommend that in developing your questionnaire you include as many potential explanatory variables as feasible to increase your chances of adequately reproducing the latent segments in an ex post analysis.

As an illustration, in this application, we use simple crosstabs to compare the PLS-POS partition with the one produced by the variables *serviceprovider* and *servicetype*. For this purpose, we suggest that you employ a statistics program such as IBM SPSS Statistics, open the extended data file (e.g., Corporate reputation data 336 pls-pos.csv), and use the crosstabs analysis option for the *serviceprovider* and PLS-POS groups variables (see Sarstedt & Mooi, 2014, for an illustration of cross tabs using IBM SPSS Statistics). Exhibit 5.17 shows the results of this analysis.

Comparing the cell counts, we find that the best match is achieved when assigning respondents who use service provider 1, 2, or 3 to PLS-POS segment 1, and those who use service provider 4 to PLS-POS segment 2. Using this grouping, $(81+71+22+25)/336=59.2\%$ of

Exhibit 5.16	Data Set and Group Information in Microsoft Excel				
cusa	*switch_1*	*switch_2*	*switch_3*	*switch_4*	**PLS-POS groups**
5	3	1	3	2	2
7	5	5	4	4	1
6	4	3	2	3	2
6	3	4	4	2	2
6	5	5	5	4	2
6	5	2	4	4	2
7	5	3	5	4	1
4	4	1	5	3	2
6	5	5	5	4	1
6	5	3	5	4	2
3	5	5	5	4	1
4	3	4	3	2	1
4	2	2	3	1	1
5	4	3	4	3	1
5	4	4	1	3	1

Exhibit 5.17	Cross Tab of PLS-POS Partition and Service Provider		
	PLS-POS segments		
Service Provider	**1**	**2**	**Sum**
1	81 (24.11%)	47 (13.99%)	**128 (38.10%)**
2	71 (21.13%)	52 (15.48%)	**123 (36.61%)**
3	22 (6.55%)	21 (6.25%)	**43 (12.80%)**
4	17 (5.06%)	25 (7.44%)	**42 (12.50%)**
Sum	**191 (56.85%)**	**145 (43.15%)**	**336**

the respondents match the PLS-POS partition. However, with service provider 4 the PLS-POS segment 2 has only 42 observations. Since a sample of 42 observations is not sufficient to reliably estimate a PLS path model for segment 2, we also assign the users of service provider 3 to segment 2 because service provider 3 has almost the same number of respondents belonging to PLS-POS segment 1 and 2. Therefore, we consider respondents who use service providers 1 or 2 as matching the first latent segment, whereas users of service providers 3 and 4 as matching the second latent segment. This grouping is also reasonable from a content point of view as service providers 1 and 2 represent two large multinational telecommunication providers while service providers 3 and 4 are smaller telecommunication providers that operate only in a few countries.

Using this grouping, $(81 + 71 + 21 + 25) / 336 = 59.2\%$ of the respondents match the PLS-POS partition. Even though the overlap is only slightly below the cutoff value of 60%, this result is not very satisfactory. Besides the low overlap, the resulting groups are highly unequal in their size, whereas FIMIX-PLS and PLS-POS provided almost equally sized groups. Alternatively, when contrasting the *servicetype* variable—that we used in the PLS-MGA in Chapter 4— with the PLS-POS partition, the result is even worse with an overlap of merely 52.7%. Since no further explanatory variables are available in the data set, we continue the illustration, using *serviceprovider* as an explanatory variable. Note that clear descriptions of identified segments is rather challenging in general.

Step 5: Estimate Segment-Specific Models

The next step in the ex post analysis is to compute the segment-specific results. For this purpose, we need to define *serviceprovider* as a grouping variable. However, *serviceprovider* has four unique values, which we need to condense into two groups. In our case, the service provider values 1 and 2 correspond to the first PLS-POS segment, while the values 3 and 4 correspond to the second PLS-POS segment. In the SmartPLS **Data View**, the **Add Data Groups** option in the menu bar allows us to define these two groups. Clicking on the **Add Data Group** button opens a dialogue box in which we can define new grouping variables. Under **Group Name**, specify the new grouping variable's name (e.g., *serviceprovider 1 + 2*) and define the values under **Group Terms**, as shown in the upper part of Exhibit 5.18.

After clicking on **OK**, SmartPLS will create a group that includes those observations where the service provider is less than 3 (i.e., 1 and 2). In a similar manner, we create a new grouping variable (e.g., *serviceprovider 3 + 4*) with all observations where the service provider is higher than 2 (i.e., 3 and 4), as shown in the lower part of Exhibit 5.18.

Having defined the groups, we can separately estimate the PLS path model for each data group. To do so, run the PLS-SEM algorithm by going to **Calculate → PLS Algorithm** in the menu bar. Alternatively, you can go to the **Modeling** window, left-click on the wheel symbol in the tool bar, and select the corresponding option in the combo box that opens. The dialogue box that opens offers the standard options for running the PLS-SEM algorithm, but has a new tab labeled **Data Groups**. Retain the default settings for the PLS-SEM algorithm, the missing values, and the weighting and click on the **Data Groups** tab. As shown in Exhibit 5.19, a menu will open, from which you can select one or more groups to be analyzed separately. Check the boxes next to *serviceprovider 1 + 2* and *serviceprovider 3 + 4* to select these two groups for the group-specific PLS path model estimations, followed by **Start Calculation**.

Exhibit 5.18	**Generate Service Provider Groups in SmartPLS**

Configure Data Group

Group Name: serviceprovider 1+2

Group Terms

Link multiple terms: ◉ AND ○ OR

serviceprovider ⌄ | is lower than ⌄ | 3.0 ⌄ | ⌄

Add Term

Configure Data Group

Group Name: serviceprovider 3+4

Group Terms

Link multiple terms: ◉ AND ○ OR

serviceprovider ⌄ | is higher than ⌄ | 2.0 ⌄ | ⌄

Add Term

Exhibit 5.19	Data Groups in the PLS-SEM Algorithm

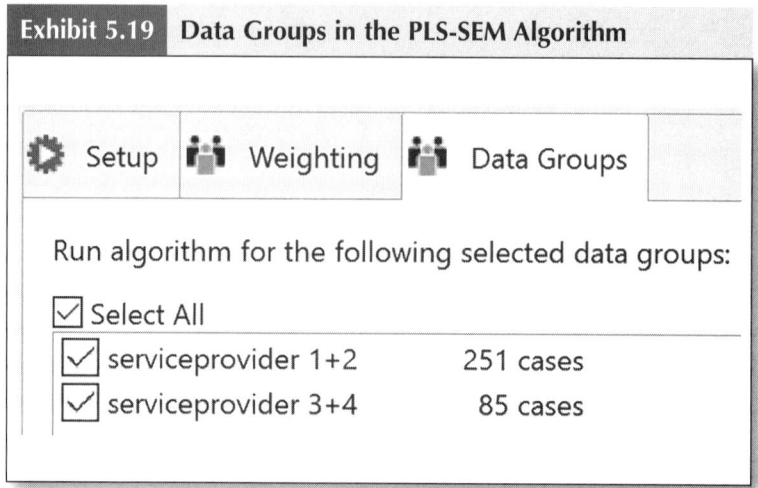

After completing the analyses of the aggregate data set and each of the groups, SmartPLS will open the results report. Initially, SmartPLS shows the results of the aggregate data analysis, but you can easily scan through the other groups by clicking on the pull-down menu next to **Data Group,** or on the **Next** button below the results tables. We also need to separately run bootstrapping for each group by going to **Calculate → Bootstrapping** or using the wheel symbol in the **Modeling** window. Use the default settings, but make certain that you again select all the groups in the **Data Groups** tab.

Exhibit 5.20 provides an overview of the aggregate data and the group-specific results, including all the reflective and formative measurement model evaluation criteria as documented in, for example, Hair et al. (2017). The measurement model evaluation results support the measures' reliability and validity.

Comparing the path coefficient estimates of the two service provider groups shows clear differences in the structural model effects. For example, whereas the effect of *COMP* on *CUSL* is not significant in the first service provider group (i.e., the large telecommunication providers), it is significantly negative in the second group (i.e., the small telecommunication providers). Similarly, the antecedents' impact on *COMP* and *LIKE* varies substantially between the groups. For example, whereas in the first group, *CSOR* has a significant effect on *LIKE*, but not on *COMP*, the opposite holds for the second group. Further analyses may involve testing whether these differences in path coefficients are significant using multigroup analysis (Chapter 4).

Exhibit 5.20	**Final Service Provider Segmentation Results**		
	Original Sample	Service Provider 1 and 2	Service Provider 3 and 4
N	336	251	85
Relative Segment Size	100.0%	74.7%	25.3%
	Path	**Path**	**Path**
ATTR → COMP	0.085	0.047	0.275***
ATTR → LIKE	0.159**	0.176**	0.001
COMP → CUSA	0.135**	0.143*	0.191
COMP → CUSL	0.011	0.097	−0.128*
CSOR → COMP	0.057	0.011	0.215**
CSOR → LIKE	0.190***	0.192***	0.105
CUSA → CUSL	0.510***	0.461***	0.611***
LIKE → CUSA	0.445***	0.501***	0.235**
LIKE → CUSL	0.334***	0.300***	0.414***
PERF → COMP	0.295***	0.357***	0.162
PERF → LIKE	0.116*	0.129	−0.017
QUAL → COMP	0.431***	0.454***	0.281**
QUAL → LIKE	0.378***	0.360***	0.664***
Reflective measures			
Convergent validity (AVE)	+	+	+
Reliability (composite reliability, Cronbach's α)	+	+	+
Discriminant validity (HTMT$_{inference}$)	+	+	+

(Continued)

Exhibit 5.20 (Continued)			
Formative measures			
Convergent validity	+	+	+
Collinearity	+	+	+
Significance and relevance of the indicators	+	+	+
R^2 values			
COMP	0.629	0.646	0.654
CUSA	0.293	0.365	0.141
CUSL	0.562	0.557	0.625
LIKE	0.557	0.570	0.525
Weighted R^2 values			
COMP	0.629	0.648	
CUSA	0.293	0.308	
CUSL	0.562	0.574	
LIKE	0.557	0.559	

***$p \leq 0.01$; **$p \leq 0.05$; *$p \leq 0.10$.

+/− = measurement model evaluation criterion fulfilled/not fulfilled in accordance with Hair et al. (2017).

However, comparing the path coefficients from the two service provider groups with those from the PLS-POS analysis shows that the results do not align well. For example, the significant but relatively weak 0.143 relationship between **COMP** and **CUSA** in service provider segment 1 is considerably higher in the PLS-POS (0.673) solution. The same holds for most of the other structural model relationships, which suggests that the PLS-POS results cannot be adequately reproduced by using the service provider variable. This result is not surprising given the limited overlap between the PLS-POS and service provider partitions.

To further assess the segmentation solution, we compute the weighted R^2 values as the sum of the segment-specific R^2 values, weighted by the relative segment size. For example, for $COMP$ in the service provider grouping, the weighted R^2 is $0.747 \cdot 0.646 + 0.253 \cdot 0.654 = 0.648$. Next, we can compare the resulting weighted R^2 values (produced by the data grouping) with the overall R^2 values resulting from the aggregate data level analysis. This analysis shows that the grouping using the service provider variables increases the model's in-sample predictive power compared to the aggregate level analysis.

SUMMARY

- **Comprehend the FIMIX-PLS and PLS-POS approaches for identifying and treating unobserved heterogeneity.** FIMIX-PLS is the first and most prominent latent class approach in PLS-SEM. Drawing on the mixture regression concept, FIMIX-PLS simultaneously estimates the path coefficients of each observation's group membership for a predefined number of groups. FIMIX-PLS proves particularly useful for identifying the number of segments to extract from the data but does not reliably identify the underlying segment structure and disregards heterogeneity in the measurement models. PLS-POS, a hill-climbing approach that gradually reallocates observations between the segments with the aim of maximizing the overall explained variance across all segments, overcomes these limitations. The approach has been shown to reliably identify segment-specific path coefficients but does not give a clear indication regarding the number of segments to consider in the analysis.

- **Understand how to use FIMIX-PLS and PLS-POS jointly in a latent class analysis.** In light of each method's strengths and weaknesses, a latent class analysis in PLS-SEM first draws on FIMIX-PLS to identify the appropriate number of segments to retain from the data. The FIMIX-PLS solution then serves as a starting solution for running PLS-POS thereafter. PLS-POS improves the FIMIX-PLS solution and allows considering heterogeneity in the measurement models of latent variables. A subsequent ex post analysis aims at identifying explanatory variables that can be used to identify a data partition into (observable) groups, which largely corresponds to the latent segment structure produced by PLS-POS. The final step involves testing segment-specific path models drawing on the segments that are defined by the previously identified explanatory variables.

- **Appreciate how to execute a FIMIX-PLS analysis using SmartPLS.** An important consideration before running FIMIX-PLS involves the treatment of missing values. In light of the considerable biases that would occur, avoid using mean value replacement. Instead, delete all observations with missing values by using casewise deletion. Running the FIMIX-PLS procedure itself requires making several choices regarding the algorithm settings. When running FIMIX-PLS, use a stop criterion of 10^{-5}, a maximum number of 5,000 iterations, and 10 repetitions to avoid convergence on a local optimum. To define a range of reasonable segment numbers, use one segment as the lower bound and the largest integer when dividing the sample size by the minimum segment sample size as the upper bound. This upper bound can be adjusted, depending on the complexity of the model and the sample size. After running FIMIX-PLS, the results include, for example, the final partition, the relative segments sizes, the segment retention criteria outcomes, and the segment-specific PLS-SEM results.

- **Know how to use SmartPLS to run a PLS-POS analysis.** Running PLS-POS in SmartPLS requires making several decisions. Researchers need to determine the number of segments to retain from the data. This decision should be based on the FIMIX-PLS results, either by specifying the segment number in line with the information and classification criteria or by using the FIMIX-PLS solution as the starting partition. When using a random initialization, researchers should run the PLS-POS algorithm 10 times to avoid convergence in local optima. In a random initialization process, researchers have the option of running a presegmentation procedure to potentially improve the PLS-POS solution. PLS-POS also requires researchers to specify the search depth, which should be equal to the number of observations. Finally, researchers can choose among different optimization criteria. Using the optimization criterion of the weighted R^2 of all endogenous constructs in the PLS path model is commonly a good approach. After running the algorithm, PLS-POS returns, among others, the final assignment of observations to the segments, the segments sizes, the optimization criterion outcome, and the segment-specific PLS-SEM results.

REVIEW QUESTIONS

1. What is the purpose of latent class analysis?

2. What is FIMIX-PLS? What are the method's strengths and weaknesses?

3. What is PLS-POS, and how does the method function?

4. Why should you jointly use FIMIX-PLS and PLS-POS in a latent class analysis?

5. What is the purpose of the ex post analysis, and why is it important?

CRITICAL THINKING QUESTIONS

1. What is the purpose of the entropy criterion, and why is its consideration important when deciding on the number of segments?

2. Under which circumstances would we negate the presence of significant unobserved heterogeneity and continue with the aggregate level analysis?

3. Why is collecting sufficient potential explanatory variables in the research design stage important for successfully running an ex post analysis?

4. Critically discuss the following statement: "As the results from an ex post analysis will never fully meet the latent segment structure produced by the latent class methods, it will necessarily have adverse consequences for the validity of results."

KEY TERMS

AIC

AIC_3

Akaike's Information Criterion (AIC)

Bayesian Information Criterion (BIC)

BIC

CAIC

Cluster analysis

Clustering

Consistent AIC (CAIC)

EN

Ex post analysis

Expectation maximization (EM) algorithm

FIMIX-PLS

Finite mixture partial least squares (FIMIX-PLS)

Information criteria

k-means clustering

Label switching

Latent class techniques

Model selection criteria

Modified AIC with factor 3 (AIC_3)

Normed entropy statistic (EN)

Observed heterogeneity

Optimization criterion

Prediction errors

Prediction-oriented segmentation in PLS-SEM (PLS-POS)

Presegmentation

Response-based segmentation techniques

Search depth

Unobserved heterogeneity

SUGGESTED READINGS

Becker, J.-M., Rai, A., Ringle, C. M., & Völckner, F. (2013). Discovering unobserved heterogeneity in structural equation models to avert validity threats. *MIS Quarterly*, 37, 665–694.

Hahn, C., Johnson, M. D., Herrmann, A., & Huber, F. (2002). Capturing customer heterogeneity using a finite mixture PLS approach. *Schmalenbach Business Review, 54,* 243–269.

Hair, J. F., Sarstedt, M., Matthews, L., & Ringle, C. M. (2016). Identifying and treating unobserved heterogeneity with FIMIX-PLS: Part I – Method. *European Business Review, 28,* 63–76.

Matthews, L., Sarstedt, M., Hair, J. F., & Ringle, C. M. (2016). Identifying and treating unobserved heterogeneity with FIMIX-PLS: Part II—A case study. *European Business Review*, 28, 208–224.

Sarstedt, M., Becker, J.-M., Ringle, C. M., & Schwaiger, M. (2011). Uncovering and treating unobserved heterogeneity with FIMIX-PLS: Which model selection criterion provides an appropriate number of segments? *Schmalenbach Business Review*, *63*, 34–62.

Wedel, M., & Kamakura, W. (2000). *Market segmentation: Conceptual and methodological foundations* (2nd ed.). Norwell, MA: Kluwer Academic.

Glossary

AIC: see *Akaike's Information Criterion.*

AIC$_3$: see *Modified AIC with factor 3.*

Akaike's Information Criterion (AIC): see *Information criteria.*

Alpha inflation: refers to the fact that the more tests you conduct at a certain significance level, the more likely you are to claim a significant result when this is not so (i.e., a Type I error).

Attribute: an element of the construct definition. It defines the general type of property to which the focal construct refers, such as an attitude.

Average variance extracted (AVE): the degree to which the construct explains the variance of its indicators.

Bandwidth-fidelity tradeoff: a practical dilemma resulting from the tradeoff between using measures that will cover the majority of variation in a trait (domain-level measurement) or measures that will assess a few specific traits (facet-level measurement) more precisely.

Bayesian Information Criterion (BIC): see *Information criteria.*

Bias-corrected and Bonferroni-adjusted confidence intervals: a confidence interval type used for testing the significance of multiple tetrads considered per measurement model.

BIC: see *Bayesian Information Criterion.*

Bonferroni correction: a method used to counteract the increase in the familywise error rate when performing multiple comparisons across several groups of data.

Bootstrap samples: make up the number of samples drawn in the bootstrapping procedure. Generally, 5,000 or more samples are recommended.

Bootstrapping: a resampling technique that draws a large number of samples from the original data (with replacement) and estimates models for each sample. It is used to determine standard errors of coefficient estimates to assess the coefficients' statistical significance without relying on distributional assumptions.

Bottom-up approach: a way to establish an HCM in which several latent variables (the LOCs) are combined into a single, more abstract construct (the HOC).

CAIC: see *Consistent AIC.*

Causal indicators: an indicator type used in formative measurement models. Constructs measured with causal indicators have an error term, which implies that the indicators do not fully form the construct.

CB-SEM: see *Covariance-based structural equation modeling.*

Cluster analysis: a class of methods that partitions a set of objects with the goal of obtaining high similarity within the formed groups and high dissimilarity between the groups.

Clustering: see *Cluster analysis.*

Collect model: an HCM type in which the HOC is a combination of several specific LOCs representing more concrete components that form the general concept. The relationship between HOC and LOCs is formative.

Common factor model: assumes that each indicator in a set of observed measures is a linear function of one or more common factors. Exploratory factor analysis (EFA), confirmatory factor analysis (CFA), and covariance-based SEM (CB-SEM) are the three main types of analyses based on common factor models.

Common variance: the variance that an indicator shares with other indicators in the measurement model of a construct.

Composite indicators: an indicator type used in formative measurement models. Constructs measured with composite indicators have no error term, which implies that the indicators fully form the construct.

Composite model approach: an approach to estimating construct proxies. Its objective is to account for the total variance in the observed indicators rather than to explain the correlations between the indicators.

Compositional invariance: exists when the composite scores are equal across groups.

Conceptual variables: broad ideas or thoughts about abstract concepts that researchers establish and propose to measure in their research.

Confidence interval: provides the lower and upper limit of values within which a true population parameter will fall with a certain probability (e.g., 95%). In PLS-SEM, the construction of the interval relies on bootstrapping standard errors.

Configural invariance: exists when constructs are equally parameterized and estimated across groups.

Confirmatory tetrad analysis in PLS-SEM (CTA-PLS): allows statistical testing of whether a measurement model is reflective or formative.

Consistent AIC (CAIC): see *Information criteria.*

Consistent PLS (PLSc): a variant of the standard PLS-SEM algorithm, which provides consistent model estimates that disattenuate the correlations between pairs of latent variables, thereby mimicking CB-SEM results.

Construct definition: the specific way in which a conceptual variable is measured in a particular study, and could differ from one study to another.

Constructs: measure concepts that are abstract, complex, and cannot be directly observed. Constructs are represented in path models as circles or ovals.

Covariance-based structural equation modeling (CB-SEM): used to confirm (or reject) theories. It does this by determining how well a proposed theoretical model can estimate the covariance matrix for a sample data set.

CTA-PLS: see *Confirmatory tetrad analysis in PLS-SEM (CTA-PLS)*.

Cubic effect: a nonlinear relationship represented by a polynomial of the degree 3; see also *nonlinear effect* and *polynomial*.

Direct effect: the direct relationship between two latent variables in a PLS path model.

Effect indicators: see *Reflective indicators*.

EN: see *Normed entropy statistic*.

Equality of composite mean values and variances: the final requirement for establishing full measurement invariance.

Equidistant scale: a scale in which the intervals are distributed in equal units. For example, a 5-point Likert scale with two negative categories (completely disagree and disagree), a neutral option, and two positive categories (agree and completely agree) can be considered an equidistant scale.

Error variance: part of an indicator's unique variance, which is assumed to be random and unreliable (i.e., measurement and sampling error).

Ex post analysis: aims at identifying one or more explanatory variable(s) that match the latent class segmentation results in the best possible way to facilitate a multigroup analysis.

Expectation maximization (EM) algorithm: an iterative algorithm for finding maximum likelihood estimates of parameters in a statistical model. The algorithm alternates between performing an expectation (E) step and a maximization (M) step. The E step creates a function for the expectation of the log-likelihood, which is evaluated using the current estimate of the parameters. The M step computes parameters by maximizing the expected log-likelihood found in the E step. The E and M steps are successively applied until the results stabilize.

Factor (score) indeterminacy: means that there is an infinite number of sets of factor scores matching the requirements of a certain common factor model.

Factor-based SEM: see *Covariance-based structural equation modeling*.

Familywise error rate: the probability of making one or more Type I errors when performing multiple comparisons across several groups of data.

FIMIX-PLS: see *Finite mixture partial least squares (FIMIX-PLS)*.

Finite mixture partial least squares (FIMIX-PLS): a latent class approach that allows for identifying and treating unobserved heterogeneity in PLS path models. The approach applies mixture regressions to simultaneously estimate group-specific parameters and observations' probabilities of segment membership.

Focal object: an element of the construct definition. It refers to the entity to which an attribute is applied.

Forced-choice scale: a scale without a neutral category.

Formative measurement model: a type of measurement model setup in which the direction of the arrows is from the indicator variables to the construct.

Formative-formative HCM: has formative measurement models of all LOCs in the HCM and formative path relationships between the LOCs and the HOC (i.e., the LOCs form the HOC).

Formative-reflective HCM: has formative measurement models of all LOCs in the HCM and reflective path relationships from the HOC to the LOCs.

Full measurement invariance: is confirmed when (1) configural invariance, (2) compositional invariance, and (3) equality of composite mean values and variances are demonstrated.

HCM: see *Hierarchical component model.*

Hierarchical common factor model: see *reflective-reflective HCM.*

Hierarchical component model (HCM): a higher-order structure (usually second-order) that contains several layers of constructs and involves a higher level of abstraction. HCMs involve a more abstract *higher-order construct (HOC)* related to two or more *lower-order constructs (LOCs)* in a reflective or formative way.

Higher-order construct (HOC): a general construct that represents all underlying LOCs in an HCM.

Higher-order model: see *Hierarchical component model (HCM).*

HOC: see *Higher-order construct.*

Impact-performance map: a graphical representation of the impact-performance map results; see *Importance-performance map analysis.*

Importance: a term used in the context of *IPMA*. It is equivalent to the unstandardized total effect of some latent variable on the target variable.

Importance-performance map analysis (IPMA): extends the standard PLS-SEM results reporting of path coefficient estimates by adding a dimension to the analysis that considers the average values of the latent variable scores. More precisely, the IPMA relates structural model total effects on a specific target construct with the average latent variable scores of this construct's predecessors.

Importance-performance matrix: see *Importance-performance map analysis (IPMA)*.

Indicators: directly measured observations (raw data), generally referred to as either *items* or *manifest variables*, represented in path models as rectangles.

Indirect effect: represents a relationship between two latent variables via a third (e.g., mediator) construct in the PLS path model. If p_1 is the relationship between a latent variable and the mediator variable, and p_2 is the relationship between the mediator variable and the endogenous latent variable, the indirect effect is the product of path p_1 and path p_2.

Information criteria: statistical measures of the relative quality of a certain segment solution that contrast the fit (i.e., the likelihood) of a solution and the number of parameters used to achieve that fit, which increase with the number of segments. The smaller the value of a certain information criterion, the better the segmentation solution. Research has proposed a variety of information criteria such as Akaike's Information Criterion (AIC), Modified AIC with factor 3 (AIC_3), Consistent AIC (CAIC), and Bayesian Information Criterion (BIC), which exhibit different tendencies to under- and overestimate the appropriate segment number.

Interaction term: an auxiliary variable entered into the path model to account for the quadratic effect, when considering self-interaction; see also *self-interaction*.

IPMA: see *Importance-performance map analysis (IPMA)*.

Items: see *Indicators*.

Jangle fallacy: describes the inference that two measures (e.g., scales) with different names measure different constructs.

k-means clustering: a group of nonhierarchical clustering algorithms that work by partitioning observations into a predefined number of groups and then iteratively reassigning observations until some numeric goal related to cluster distinctiveness is met.

Label switching: occurs when the label of a specific segment changes from one FIMIX-PLS run to the next.

Latent class techniques: a class of approaches that facilitates uncovering unobserved heterogeneity. Different approaches have been proposed, which draw on, for example, finite mixture, genetic algorithm, or hill-climbing approaches to PLS-SEM.

Latent variable: see *Constructs*.

Latent variable scores: the values calculated to represent latent variables.

Linear effect: represented by a straight line when plotted as a graph.

Linear relationship: see *Linear effect*.

LOC: see *Lower-order construct.*

Log transformation: a type of data transformation to account for nonlinear relationships, which applies a base 10 logarithm to every observation.

Lower-order construct (LOC): a subdimension of the HOC in an HCM.

Manifest variables: see *Indicators.*

Measurement equivalence: see *Measurement invariance.*

Measurement invariance: deals with the comparability of responses to sets of items across groups. Among other things, measurement invariance implies that the categorical moderator variable's effect is restricted to the path coefficients and does not involve group-related differences in the measurement models.

Measurement invariance of composite models (MICOM) procedure: a series of tests to assess invariance of measures (constructs) across multiple groups of data. The procedure comprises three steps that test different aspects of measurement invariance: (1) configural invariance (i.e., equal parameterization and way of estimation), (2) compositional invariance (i.e., equal indicator weights), and (3) equality of composite mean values and variances.

Measurement model: an element of a path model that contains the indicators and their relationships with the constructs.

Measurement model misspecification: describes the use of a reflective measurement model when it should be formative or the use of a formative measurement model when it should be reflective. Measurement model misspecification usually yields invalid results and misleading conclusions.

Mediation: represents a situation in which a mediator variable to some extent absorbs the effect of a latent variable on an endogenous latent variable in the PLS path model.

Mediator model: see *Mediation.*

Metric scale: a type of measurement scale that has a constant unit of measurement so that the distance between the scale points is equal (interval scale), thereby allowing for the interpretation of the scale points' absolute differences. In case the scale additionally has an absolute zero point, ratios among the scale points can be interpreted (ratio scale).

MICOM: see *Measurement invariance of composite models (MICOM) procedure.*

MIMIC model: See *Multiple indicators and multiple causes model.*

Model selection criteria: see *Information criteria.*

Moderation: occurs when the effect of a latent variable on an endogenous latent variable depends on the values of a third variable, referred to as a moderator variable.

Modified AIC with factor 3 (AIC_3): see *Information criteria.*

Multigroup analysis: tests whether parameters (mostly path coefficients) differ significantly between two groups. Research has proposed a range of approaches to multigroup analysis, which rely on the bootstrapping or permutation procedure.

Multiple battery model: an HCM type in which the LOCs measure the same construct at different points of time using data from the same set of respondents.

Multiple indicators and multiple causes (MIMIC) model: a type of structural equation model that incorporates both formative and reflective indicators to measure latent constructs.

Multiple testing problem: occurs when the Type I error of a series of tests increases exponentially.

Nominal scale: a measurement scale in which numbers are assigned that can be used to identify and classify objects (e.g., people, companies, products).

Nonlinear effect: is not represented by a straight line when plotted on a graph but by a curve.

Nonlinear relationship: see *Nonlinear effect.*

Nonredundant tetrads: tetrads considered for significance testing in CTA-PLS.

Normed entropy statistic (EN): a statistical measure that uses the observations' probabilities of segment membership as an indication of how well segments are separated. Values above 0.50 permit a clear-cut classification of data into the segments.

Observed heterogeneity: occurs when the sources of heterogeneity are known and can be traced back to observable characteristics such as demographics (e.g., gender, age, income).

Omnibus test of group differences (OTG): a PLS-SEM-based multigroup analysis method that compares the parameter results of two or more groups at the same time. It corresponds to the F test in regression or ANOVA.

Optimization criterion: a certain measure whose value defines the quality of a tested set of parameters.

Ordinal scale: a measurement scale in which the assigned numbers indicate the relative positions of objects in an ordered series.

Orthogonalizing approach: an approach to model the nonlinear (e.g., quadratic) term with orthogonal indicators.

OTG: see *Omnibus test of group differences (OTG).*

Parametric test: a multigroup variant, representing a modified version of a two-independent-samples *t* test when the population variances of the groups are considered to be equal (homoscedastic). Otherwise (i.e., the variances are unequal or heteroscedastic), the parametric test uses a modified version of the Welch-Satterthwaite test.

Partial least squares algorithm: allows estimating path models with latent variables.

Partial least squares path modeling: see *Partial least squares structural equation modeling.*

Partial least squares regression (PLS-R): an approach designed to reduce the problem of multicollinearity in regression models. It uses a principal components analysis that extracts linear composites of the independent variables and their respective scores, taking into consideration the relationship between the independent and dependent variables, and maximizing the explanation of the dependent variable.

Partial least squares structural equation modeling (PLS-SEM): a composite-based method to estimate structural equation models. The goal is to maximize the explained variance of the endogenous latent variables.

Partial measurement invariance: is confirmed when only (1) configural invariance and (2) compositional invariance are demonstrated.

Performance: the term used in the context of *IPMA* to represent the mean value of the unstandardized (and rescaled) scores of a latent variable or an indicator.

Permutation: generates a reference distribution of some set of parameters from the actual data by randomly exchanging observations between the groups multiple times.

Permutation test: a multigroup analysis variant in PLS-SEM. It randomly permutes observations between the groups and re-estimates the model to derive a test statistic for the group differences.

PLS algorithm: see *Partial least squares algorithm.*

PLSc: see *Consistent PLS.*

PLS-MGA: a bootstrap-based multigroup analysis technique.

PLS-POS: see *Prediction-oriented segmentation approach in PLS-SEM (PLS-POS).*

PLS-R: see *Partial least squares regression.*

PLS-SEM: see *Partial least squares structural equation modeling.*

Polynomial: a mathematical expression consisting of a sum of terms, whereby each term includes a variable raised to a power and multiplied by a coefficient.

Polynomial degree: determines number of terms that are summed in a polynomial; see also *polynomial.*

Prediction errors: are used in PLS-POS to reassign observations from one segment to the other. They correspond to the observation's squared differences between the predicted and actual scores of the endogenous latent variable(s).

Prediction-oriented segmentation in PLS-SEM (PLS-POS): a distance-based segmentation method for PLS-SEM.

Premature convergence: occurs when the PLS-SEM algorithm converges but the results depend on the starting values and are therefore not optimal.

Presegmentation: an option for PLS-POS that, in the first round, assigns all observations at the same time to their closest segment. Then in the subsequent iterations PLS-POS reassigns only one observation per iteration.

Principal components regression: performs a principal components analysis on the independent variables, and the principal components are used as predictive/explanatory variables for the dependent variable. It focuses on reducing the dimensionality of the independent variables without taking into account the relationship between the independent and dependent variables.

Priority map analysis: see *Importance-performance map analysis (IPMA)*.

Product indicator approach: an approach to model the nonlinear (e.g., quadratic) term. It involves multiplying all indicators of the exogenous latent variable to establish a measurement model of the nonlinear (e.g., quadratic) term. The approach is only applicable for reflectively measured exogenous latent variables.

Product indicators: indicators of an interaction term, generated by multiplication of each indicator of a construct with each indicator of the moderator variable; see also *product indicator approach*.

Quadratic effect: represented by a curved nonlinear relationship characterized by a polynomial of the degree 2; see also *nonlinear effect* and *polynomial*.

Reflective indicators: indicators of a *reflective measurement model*.

Reflective measurement model: a type of measurement model setup in which the direction of the arrows is from the construct to the indicator variables, indicating the assumption that the construct causes the measurement (more precisely, the covariation) of the indicator variables. Indicators in reflective measurement models are also referred to as effect indicators.

Reflective-formative HCM: has reflective measurement models of all LOCs in the HCM and formative path relationships from the LOCs to the HOC.

Reflective-reflective HCM: has reflective measurement models of all LOCs in the HCM and reflective path relationships from the HOC to the LOCs; also referred to as *hierarchical common factor model*.

Repeated indicators approach for HCM: a type of measurement model setup in HCM that uses the indicators of the LOCs as indicators of the HOC to create an HCM in PLS-SEM.

Rescaling: changes the values of a variable's scale to fit a predefined range (e.g., 0 to 100).

Response-based segmentation techniques: see *Latent class techniques*.

Search depth: a parameter in PLS-POS that defines the maximum number of observations considered for reassignment to another segment.

Second-order construct: see *Hierarchical component model.*

Self-interaction: occurs when the effect of an exogenous latent variable on an endogenous latent variable depends on the values of the exogenous latent variable.

Šidák procedure: a method used to counteract the increase in the family-wise error rate when performing multiple comparisons across several groups of data.

Specific variance: part of an indicator's unique variance, which is assumed to be systematic and reliable.

Spread model: an HCM type in which the HOC is manifested in several more specific LOCs. The relationship between HOC and LOCs is reflective.

Stand-alone HCM: an HCM that is not embedded within a greater nomological net of constructs.

Tetrad (τ): the difference of the product of a pair of covariances and the product of another pair of covariances. In reflective measurement models, this difference is assumed to be zero or at least close to zero; that is, the tetrad is expected to vanish. Nonvanishing tetrads in a latent variable's measurement model cast doubt on its reflective specification, suggesting a formative specification.

Theoretical model: a set of equations with variables that formalize a theory.

Top-down approach: a way to establish an HCM in which a more abstract construct (the HOC) is defined that consists of several components (the LOCs).

Total effect: the sum of the direct effect and the indirect effect(s) between a latent variable and an endogenous latent variable in the PLS path model.

Total effects analysis of collect-type HCMs: links the antecedent construct of an HOC in a reflective-formative or formative-formative HCM with the LOCs. The effect of the antecedent construct on the HOC is equal to its direct effect plus the sum of indirect effects via the LOCs.

Two-stage approach (nonlinear effects): an approach to model the nonlinear (e.g., quadratic) term. The approach can be used for any kind of exogenous construct no matter whether it is measured reflectively, formatively, or represents a single item construct.

Two-stage approach for HCMs: allows estimating path relationships between an antecedent construct of an HOC in a *reflective-formative HCM or a formative-formative HCM.*

Unique variance: the variance an indicator does not share with other indicators in the measurement model of a construct. It consists of specific variance and error variance.

Unobserved heterogeneity: occurs when the sources of heterogeneous data structures are not (fully) known.

Vanishing tetrad: see *Tetrad*.

References

Aguinis, H., Beaty, J. C., Boik, R. J., & Pierce, C. A. (2005). Effect size and power in assessing moderating effects of categorical variables using multiple regression: A 30-year review. *Journal of Applied Psychology, 90*, 94–107.

Akaike, H. (1973). Information theory and an extension of the maximum likelihood principle. In B. N. Petrov & F. Csaki (Eds.), *Second international symposium on information theory* (pp. 267–281). Budapest, Hungary: Academiai Kiadó.

Al-Gahtani, S. S., Hubona, G. S., & Wang, J. (2007). Information technology (IT) in Saudi Arabia: Culture and the acceptance and use of IT. *Information & Management, 44*, 681–691.

Albers, S. (2010). PLS and success factor studies in marketing. In V. Esposito Vinzi, W. W. Chin, J. Henseler, & H. Wang (Eds.), *Handbook of partial least squares: Concepts, methods and applications in marketing and related fields* (pp. 409–425). Berlin, Heidelberg: Springer.

Anderberg, M. R. (1973). *Cluster analysis for applications.* New York, NY: Academic Press.

Antonakis, J., Bendahan, S., Jacquart, P., & Lalive, R. (2010). On making causal claims: A review and recommendations. *Leadership Quarterly, 21*, 1086–1120.

Astrachan, C. B., Patel, V. K., & Wanzenried, G. (2014). A comparative study of CB-SEM and PLS-SEM for theory development in family firm research. *Journal of Family Business Strategy, 5*, 116–128.

Atinc, G., Simmering, M. J., & Kroll, M. J. (2012). Control variable use and reporting in macro and micro management research. *Organizational Research Methods, 15*, 57–74.

Babin, B. J., Hair, J. F., & Boles, J. S. (2008). Publishing research in marketing journals using structural equation modelling. *Journal of Marketing Theory and Practice, 16*, 279–286.

Barroso, C., & Picón, A. (2012). Multi-dimensional analysis of perceived switching costs. *Industrial Marketing Management, 41*, 531–543.

Bass, F. M. (1969). A new product growth for model consumer durables. *Management Science, 15*, 215–227.

Bearden, W. O., Netemeyer, R. G., & Haws, K. L. (2011). *Handbook of marketing scales: Multi-item measures of marketing and consumer behavior research.* Thousand Oaks, CA: Sage.

Becker, J.-M., Klein, K., & Wetzels, M. (2012). Hierarchical latent variable models in PLS-SEM: Guidelines for using reflective-formative type models. *Long Range Planning, 45*, 359–394.

Becker, J.-M., Rai, A., & Rigdon, E. E. (2013). Predictive validity and formative measurement in structural equation modeling: Embracing practical relevance. In *Proceedings of the 34th International Conference on Information Systems*.

Becker, J.-M., Rai, A., Ringle, C. M., & Völckner, F. (2013). Discovering unobserved heterogeneity in structural equation models to avert validity threats. *MIS Quarterly, 37,* 665–694.

Becker, J.-M., Ringle, C. M., Sarstedt, M., & Völckner, F. (2015). How collinearity affects mixture regression results. *Marketing Letters, 26,* 643–659.

Bentler, P. M., & Huang, W. (2014). On components, latent variables, PLS and simple methods: Reactions to Rigdon's rethinking of PLS. *Long Range Planning, 47,* 136–145.

Binz Astrachan, C., Patel, V. K., & Wanzenried, G. (2014). A comparative study of CB-SEM and PLS-SEM for theory development in family firm research. *Journal of Family Business Strategy, 5,* 116–128.

Bollen, K. A. (1989). *Structural equations with latent variables*. New York, NY: Wiley.

Bollen, K. A. (2002). Latent variables in psychology and the social sciences. *Annual Review of Psychology, 53,* 605–634.

Bollen, K. A. (2011). Evaluating effect, composite, and causal indicators in structural equation models. *MIS Quarterly, 35,* 359–372.

Bollen, K. A., & Bauldry, S. (2011). Three Cs in measurement models: Causal indicators, composite indicators, and covariates. *Psychological Methods, 16,* 265–284.

Bollen, K. A., & Davis, W. R. (2009). Causal indicator models: Identification, estimation, and testing. *Structural Equation Modeling, 16,* 498–522.

Bollen, K. A., & Diamantopoulos, A. (2017). In defense of causal-formative indicators: A minority report. *Psychological Methods*. Advance online publication. doi:10.1037/met0000056

Bollen, K. A., & Ting, K.-F. (1993). Confirmatory tetrad analysis. In P. V. Marsden (Ed.), *Sociological methodology* (pp. 147–175). Washington, DC: American Sociological Association.

Bollen, K. A., & Ting, K.-F. (2000). A tetrad test for causal indicators. *Psychological Methods, 5,* 3–22.

Borsboom, D., Mellenbergh, G. J., & van Heerden, J. (2003). The theoretical status of latent variables. *Psychological Review, 110,* 203–219.

Bozdogan, H. (1987). Model selection and Akaike's Information Criterion (AIC): The general theory and its analytical extensions. *Psychometrika, 52,* 345–370.

Bozdogan, H. (1994). Mixture-model cluster analysis using model selection criteria in a new information measure of complexity. In H. Bozdogan (Ed.), *Proceedings of the First US/Japan conference on frontiers of statistical modelling: An information approach* (Vol. 2, pp. 69–113). Boston, MA: Kluwer Academic.

Bruner, G. C., James, K. E., & Hensel, P. J. (2001). *Marketing scales handbook: A compilation of multi-item measures*. Oxford, UK: Butterworth-Heinemann.

Campbell, D. T., & Stanley, J. C. (1966). *Experimental and quasi-experimental designs for research*. Boston, MA: Houghton Mifflin.

Carrión, G. C., Henseler, J., Ringle, C. M., & Roldán, J. L. (2016). Prediction-oriented modeling in business research by means of PLS path modeling: Introduction to a JBR special section. *Journal of Business Research, 69*, 4545–4551.

Cassel, C., Hackl, P., & Westlund, A. H. (1999). Robustness of partial least squares method for estimating latent variable quality structures. *Journal of Applied Statistics, 26*, 435–446.

Chin, W. W. (1994). *PLS Graph* [Computer software]. Calgary, Canada: University of Calgary.

Chin, W. W. (1998). The partial least squares approach to structural equation modeling. In G. A. Marcoulides (Ed.), *Modern methods for business research* (pp. 295–358). Mahwah, NJ: Lawrence Erlbaum.

Chin, W. W. (2003). A permutation procedure for multigroup comparison of PLS models. In M. Vilares, M. Tenenhaus, P. Colho, V. Esposito Vinzi, & A. Morineau (Eds.), *PLS and related methods: Proceedings of the International Symposium PLS'03* (pp. 33–43). Lisbon, Portugal: Decisia.

Chin, W. W. (2010). How to write up and report PLS analyses. In V. Esposito Vinzi, W. W. Chin, J. Henseler, & H. Wang (Eds.), *Handbook of partial least squares: Concepts, methods and applications in marketing and related fields* (pp. 655–690). Berlin, Heidelberg: Springer.

Chin, W. W., & Dibbern, J. (2010). A permutation based procedure for multi-group PLS analysis: Results of tests of differences on simulated data and a cross cultural analysis of the sourcing of information system services between Germany and the USA. In V. Esposito Vinzi, W. W. Chin, J. Henseler, & H. Wang (Eds.), *Handbook of partial least squares: Concepts, methods and applications in marketing and related fields* (pp. 171–193). Berlin, Germany: Springer.

Cho, H. C., & Abe, S. (2012). Is two-tailed testing for directional research hypotheses tests legitimate? *Journal of Business Research, 66*, 1261–1266.

Cliff, N. (1983). Some cautions concerning the application of causal modeling methods. *Multivariate Behavioral Research, 18*, 115–126.

Cohen, J. (1988). *Statistical power analysis for the behavioral sciences*. Mahwah, NJ: Lawrence Erlbaum.

DeVellis, R. F. (2011). *Scale development*. Thousand Oaks, CA: Sage.

Diamantopoulos, A. (1994). Modelling with LISREL: A guide for the uninitiated. *Journal of Marketing Management, 10*, 105–136.

Diamantopoulos, A. (2005). The C-OAR-SE procedure for scale development in marketing: A comment. *International Journal of Research in Marketing, 22*, 1–9.

Diamantopoulos, A. (2006). The error term in formative measurement models: Interpretation and modeling implications. *Journal of Modelling in Management, 1,* 7–17.

Diamantopoulos, A., & Papadopoulos, N. (2010). Assessing the cross-national invariance of formative measures: Guidelines for international business researchers. *Journal of International Business Studies, 61,* 1203–1218.

Diamantopoulos, A., & Riefler, P. (2011). Using formative measures in international marketing models: A cautionary tale using consumer animosity as an example. *Advances in International Marketing, 10,* 11–30.

Diamantopoulos, A., Riefler, P., & Roth, K. P. (2008). Advancing formative measurement models. *Journal of Business Research, 61,* 1203–1218.

Diamantopoulos, A., Sarstedt, M., Fuchs, C., Kaiser, S., & Wilczynski, P. (2012). Guidelines for choosing between multi-item and single-item scales for construct measurement: A predictive validity perspective. *Journal of the Academy of Marketing Science, 40,* 434–449.

Diamantopoulos, A., & Siguaw, J. A. (2006). Formative vs. reflective indicators in measure development: Does the choice of indicators matter? *British Journal of Management, 13,* 263–282.

Diamantopoulos, A., & Winklhofer, H. M. (2001). Index construction with formative indicators: An alternative to scale development. *Journal of Marketing Research, 38,* 269–277.

Dibbern, J., & Chin, W. W. (2005). Multi-group comparison: Testing a PLS model on the sourcing of application software services across Germany and the USA using a permutation based algorithm. In F. W. Bliemel, A. Eggert, G. Fassott, & J. Henseler (Eds.), *Handbuch PLS-Pfadmodellierung. Methode, Anwendung, Praxisbeispiele* (pp. 135–160). Stuttgart, Germany: Schäffer-Poeschel.

Dijkstra, T. K. (2014). PLS' janus face—Response to Professor Rigdon's "Rethinking partial least squares modeling: In praise of simple methods." *Long Range Planning, 47,* 146–153.

Dijkstra, T. K., & Henseler, J. (2011). Linear indices in nonlinear structural equation models: Best fitting proper indices and other composites. *Quality & Quantity, 45,* 1505–1518.

Dijkstra, T. K., & Henseler, J. (2015a). Consistent and asymptotically normal PLS estimators for linear structural equations. *Computational Statistics & Data Analysis, 81,* 10–23.

Dijkstra, T. K., & Henseler, J. (2015b). Consistent partial least squares path modeling. *MIS Quarterly, 39,* 297–316.

Dijkstra, T. K., & Schmermelleh-Engel, K. (2014). Consistent partial least squares for nonlinear structural equation models. *Psychometrika, 79,* 585–604.

Do Valle, P. O., & Assaker, G. (2016). Using partial least squares structural equation modeling in tourism research: A review of past research and recommendations for future applications. *Journal of Travel Research, 55,* 695–708.

Eberl, M. (2010). An application of PLS in multi-group analysis: The need for differentiated corporate-level marketing in the mobile communications industry. In V. Esposito Vinzi, W. W. Chin, J. Henseler, & H. Wang (Eds.), *Handbook of partial least squares: Concepts, methods and applications in marketing and related fields* (pp. 487–514). Berlin, Germany: Springer.

Eberl, M., & Schwaiger, M. (2005). Corporate reputation: Disentangling the effects on financial performance. *European Journal of Marketing, 39,* 838–854.

Edwards, J. R. (2001). Multidimensional constructs in organizational behavior research: An integrative analytical framework. *Organizational Research Methods, 4,* 144–192.

Eisenbeiss, M., Cornelißen, M., Backhaus, K., & Hoyer, W. D. (2014). Nonlinear and asymmetric returns on customer satisfaction: Do they vary across situations and consumers? *Journal of the Academy of Marketing Science, 42,* 242–263.

Ernst, M. D. (2004). Permutation methods: A basis for exact inference. *Statistical Science, 19,* 676–685.

Esposito Vinzi, V., Trinchera, L., Squillacciotti, S., & Tenenhaus, M. (2008). REBUS-PLS: A response-based procedure for detecting unit segments in PLS path modelling. *Applied Stochastic Models in Business and Industry, 24,* 439–458.

Falk, R. F., & Miller, N. B. (1992). *A primer for soft modeling.* Akron, OH: University of Akron Press.

Finn, A. (2012). Customer delight: Distinct construct or zone of nonlinear response to customer satisfaction? *Journal of Service Research, 15,* 99–110.

Fisher, R. A. (1935). *The design of experiments.* New York, NY: Hafner.

Fornell, C. G., & Bookstein, F. L. (1982). Two structural equation models: LISREL and PLS applied to consumer exit-voice theory. *Journal of Marketing Research, 19,* 440–452.

Fornell, C. G., Johnson, M. D., Anderson, E. W., Cha, J., & Bryant, B. E. (1996). The American Customer Satisfaction Index: Nature, purpose, and findings. *Journal of Marketing, 60,* 7–18.

Fu, J.-R. (2006). *VisualPLS—An enhanced GUI for LVPLS (PLS 1.8 PC)* (version 1.04) [Computer software]. Retrieved from http://www2 .kuas.edu.tw/prof/fred/vpls

Gilliam, D. A., & Voss, K. (2013). A proposed procedure for construct definition in marketing. *European Journal of Marketing, 47,* 5–26.

Good, P. (2000). *Permutation tests: A practical guide to resampling methods for testing hypotheses.* New York, NY: Springer.

Grace, J. B., & Bollen, K. A. (2007). Representing general theoretical concepts in structural equation models: The role of composite variables. *Environmental and Ecological Statistics, 15,* 191–213.

Gudergan, S. P., Ringle, C. M., Wende, S., & Will, A. (2008). Confirmatory tetrad analysis in PLS path modeling. *Journal of Business Research, 61,* 1238–1249.

Guide, V. D. R., & Ketokivi, M. (2015). Notes from the editors: Redefining some methodological criteria for the journal. *Journal of Operations Management, 37,* v–viii.

Guttman, L. (1955). The determinacy of factor score matrices with implications for five other basic problems of common-factor theory. *British Journal of Statistical Psychology, 8,* 65–81.

Hahn, C., Johnson, M. D., Herrmann, A., & Huber, F. (2002). Capturing customer heterogeneity using a finite mixture PLS approach. *Schmalenbach Business Review, 54,* 243–269.

Hair, J. F., Black, W. C., Babin, B. J., & Anderson, R. E. (2010). *Multivariate data analysis.* Englewood Cliffs, NJ: Prentice Hall.

Hair, J. F., Hollingsworth, C. L., Randolph, A. B., & Chong, A. Y. L. (2017). An updated and expanded assessment of PLS-SEM in information systems research. *Industrial Management & Data Systems.* Advance online publication. doi: 10.1108/IMDS-04-2016-0130

Hair, J. F., Hult, G. T. M., Ringle, C. M., & Sarstedt, M. (2017). *A primer on partial least squares structural equation modeling (PLS-SEM)* (2nd ed.). Thousand Oaks, CA: Sage.

Hair, J. F., Ringle, C. M., & Sarstedt, M. (2011). PLS-SEM: Indeed a silver bullet. *Journal of Marketing Theory and Practice, 19,* 139–151.

Hair, J. F., Ringle, C. M., & Sarstedt, M. (2013). Partial least squares structural equation modeling: Rigorous applications, better results and higher acceptance. *Long Range Planning, 46,* 1–12.

Hair, J. F., Sarstedt, M., Matthews, L., & Ringle, C. M. (2016). Identifying and treating unobserved heterogeneity with FIMIX-PLS: Part I— method. *European Business Review, 28,* 63–76.

Hair, J. F., Sarstedt, M., Pieper, T., & Ringle, C. M. (2012). The use of partial least squares structural equation modeling in strategic management research: A review of past practices and recommendations for future applications. *Long Range Planning, 45,* 320–340.

Hair, J. F., Sarstedt, M., Ringle, C. M., & Mena, J. A. (2012). An assessment of the use of partial least squares structural equation modeling in marketing research. *Journal of the Academy of Marketing Science, 40,* 414–433.

Hair, J. F., Hult, G. T. M., Ringle, C. M., Sarstedt, M., & Thiele, K. O. (2017). Mirror, mirror on the wall. A comparative evaluation of component-based structural equation modeling methods. *Journal of the Academy of Marketing Science.* Advance online publication. doi: 10.1007/s11747-017-0517-x

Hay, D. A., & Morris, D. J. (1991). *Industrial economics and organization: Theory and evidence.* New York, NY: Oxford University Press.

Helm, S., Eggert, A., & Garnefeld, I. (2010). Modelling the impact of corporate reputation on customer satisfaction and loyalty using PLS. In V. Esposito Vinzi, W. W. Chin, J. Henseler, & H. Wang (Eds.), *Handbook of partial least squares: Concepts, methods and applications in marketing and related fields* (pp. 515–534). Berlin, Heidelberg: Springer.

Henseler, J., & Chin, W. W. (2010). A comparison of approaches for the analysis of interaction effects between latent variables using partial least squares path modeling. *Structural Equation Modeling, 17,* 82–109.

Henseler, J., & Dijkstra, T. (2016). *ADANCO 2* [Computer software]. Retrieved from http://www.compositemodeling.com

Henseler, J., Dijkstra, T. K., Sarstedt, M., Ringle, C. M., Diamantopoulos, A., Straub, D. W., . . . Calantone, R. J. (2014). Common beliefs and reality about partial least squares: Comments on Rönkkö & Evermann (2013). *Organizational Research Methods, 17,* 182–209.

Henseler, J., Fassott, G., Dijkstra, T. K., & Wilson, B. (2012). Analysing quadratic effects of formative constructs by means of variance-based structural equation modelling. *European Journal of Information Systems, 21,* 99–112.

Henseler, J., Hubona, G., & Ray, P. A. (2016). Using PLS path modeling in new technology research: Updated guidelines. *Industrial Management & Data Systems, 116,* 2–20.

Henseler, J., Ringle, C. M., & Sarstedt, M. (2012). Using partial least squares path modeling in international advertising research: Basic concepts and recent issues. In S. Okazaki (Ed.), *Handbook of research in international advertising* (pp. 252–276). Cheltenham, UK: Edward Elgar.

Henseler, J., Ringle, C. M., & Sarstedt, M. (2015). A new criterion for assessing discriminant validity in variance-based structural equation modeling. *Journal of the Academy of Marketing Science, 43,* 115–135.

Henseler, J., Ringle, C. M., & Sarstedt, M. (2016). Testing measurement invariance of composites using partial least squares. *International Marketing Review, 33,* 405–431.

Henseler, J., Ringle, C. M., & Sinkovics, R. R. (2009). The use of partial least squares path modeling in international marketing. *Advances in International Marketing, 20,* 277–320.

Hochberg, Y., & Tamhane, A. C. (2011). *Multiple comparison procedures.* New York, NY: Wiley.

Höck, C., Ringle, C. M., & Sarstedt, M. (2010). Management of multipurpose stadiums: Importance and performance measurement of service interfaces. *International Journal of Services Technology and Management, 14,* 188–207.

Howell, R. D., Breivik, E., & Wilcox, J. B. (2007). Reconsidering formative measurement. *Psychological Methods, 12,* 205–218.

Howell, R. D., Breivik, E., & Wilcox, J. B. (2013). Formative measurement: A critical perspective. *Data Base for Advances in Information Systems, 44,* 44–55.

Hsu, H. M., Chang, I. C., & Lai, T. W. (2016). Physicians' perspectives of adopting computer-assisted navigation in orthopedic surgery. *International Journal of Medical Informatics, 94,* 207–214.

Hubona, G. (2015). *PLS-GUI* [Computer software]. Retrieved from http://www.pls-gui.com

Hult, G. T. M., Ketchen, D. J., Griffith, D. A., Finnegan, C. A., Gonzalez-Padron, T., Harmancioglu, N., . . . Cavusgil, S. T. (2008). Data equivalence in cross-cultural international business research: Assessment and guidelines. *Journal of International Business Studies, 39,* 1027–1044.

Jarvis, C. B., MacKenzie, S. B., & Podsakoff, P. M. (2003). A critical review of construct indicators and measurement model misspecification in marketing and consumer research. *Journal of Consumer Research, 30,* 199–218.

Jedidi, K., Jagpal, H. S., & DeSarbo, W. S. (1997). Finite-mixture structural equation models for response-based segmentation and unobserved heterogeneity. *Marketing Science, 16,* 39–59.

Jisha, P. R., & Thomas, I. (2016). Quality of life and infertility: Influence of gender, years of marital life, resilience, and anxiety. *Psychological Studies, 61,* 159–169.

Johnson, R. E., Rosen, C. C., & Chang, C.-H. (2011). To aggregate or not to aggregate: Steps for developing and validating higher-order multidimensional constructs. *Journal of Business Psychology, 26,* 241–248.

Jones, M. A., Mothersbaugh, D. L., & Beatty, S. E. (2000). Switching barriers and repurchase intentions in services. *Journal of Retailing, 76,* 259–274.

Jöreskog, K. G. (1969). A general approach to confirmatory maximum likelihood factor analysis. *Psychometrika, 34,* 183–202.

Jöreskog, K. G. (1973). A general method for estimating a linear structural equation system. In A. S. Goldberger & O. D. Duncan (Eds.), *Structural equation models in the social sciences* (pp. 85–112). London, UK: Academic Press.

Jöreskog, K. G. (1978). Structural analysis of covariance and correlation matrices. *Psychometrika, 43,* 443–477.

Jöreskog, K. G., & Wold, H. (1982). The ML and PLS techniques for modeling with latent variables: Historical and comparative aspects. In H. Wold & K. G. Jöreskog (Eds.), *Systems under indirect observation, Part I* (pp. 263–270). Amsterdam, Netherlands: North-Holland.

Kansky R., Kidd, M., & Knight, A. T. (2016). A wildlife tolerance model and case study for understanding human wildlife conflicts. *Biological Conservation, 201,* 137–145.

Kaufmann, L., & Gaeckler, J. (2015). A structure review of partial least squares in supply chain management research. *Journal of Purchasing & Supply Management, 21,* 259–272.

Keil, M., Saarinen, T., Tan, B. C. Y., Tuunainen, V., Wassenaar, A., & Wei, K.-K. (2000). A cross-cultural study on escalation of commitment behavior in software projects. *MIS Quarterly, 24,* 299–325.

Kenny, D. A. (2015). *Moderation.* Retrieved from http://davidakenny.net/cm/moderation.htm

Kessel, F., Ringle, C. M., & Sarstedt, M. (2010). On the impact of missing values on model selection in FIMIX-PLS. *Proceedings of the 2010 INFORMS Marketing Science Conference,* Cologne, Germany.

Kline, R. B. (2015). *Principles and practice of structural equation modeling* (4th ed.). New York, NY: Guilford Press.

Kristensen, K., Martensen, A., & Grønholdt, L. (2000). Customer satisfaction measurement at Post Denmark: Results of application of the European customer satisfaction index methodology. *Total Quality Management, 11,* 1007–1015.

Kwofie, T. E., Adinyira, E., & Fugar, F. (2015). Nature of communication ineffectiveness inherent in the procurement systems on mass housing projects. *Journal of Construction Engineering*, 914520.

Law, K. S., & Wong, C.-S. (1999). Multidimensional constructs in structural equation analysis: An illustration using the job perception and job satisfaction constructs. *Journal of Management, 25,* 143–160.

Law, K. S., Wong, C.-S., & Mobley, W. H. (1998). Toward a taxonomy of multidimensional constructs. *The Academy of Management Review, 23,* 741–755.

Lee, L., Petter, S., Fayard, D., & Robinson, S. (2011). On the use of partial least squares path modeling in accounting research. *International Journal of Accounting Information Systems, 12,* 305–328.

Lee, N., & Cadogan, J. W. (2013). Problems with formative and higher-order reflective variables. *Journal of Business Research, 66,* 242–247.

Liang, Z., Jaszak, R. J., & Coleman, R. E. (1992). Parameter estimation of finite mixtures using the EM algorithm and information criteria with applications to medical image processing. *IEEE Transactions on Nuclear Science, 39,* 1126–1133.

Liu, J., Zhao, X., & Yan, P. (2016). Risk paths in international construction projects: Case study from Chinese contractors. *Journal of Construction Engineering and Management, 142,* http://dx.doi.org/10.1061/(ASCE) CO.1943-7862.0001116

Lohmöller, J.-B. (1984). Das programmsystem LVPLS für pfadmodelle mit latenten variablen [The program system LVPLS for path models with latent variables]. *ZA-Information/Zentralarchiv für Empirische Sozialforschung, 14,* 44–51.

Lohmöller, J.-B. (1987). *LVPLS 1.8.* Cologne, Germany: Zentralarchiv für Empirische Sozialforschung.

Lohmöller, J.-B. (1989). *Latent variable path modeling with partial least squares.* Heidelberg, Germany: Physica.

MacCallum, R. C., & Browne, M. W. (1993). The use of causal indicators in covariance structure models: Some practical issues. *Psychological Bulletin, 114,* 533–541.

MacCallum, R. C., Browne, M. W., & Cai, L. (2007). Factor analysis models as approximations. In R. Cudeck & R. C. MacCallum (Eds.), *Factor analysis at 100: Historical developments and future directions* (pp. 153–175). Mahwah, NJ: Lawrence Erlbaum.

MacKenzie, S. B., Podsakoff, P. M., & Podsakoff, N. P. (2011). Construct measurement and validation procedures in MIS and behavioral research: Integrating new and existing techniques. *MIS Quarterly, 35,* 293–334.

Marcoulides, G. A., & Chin, W. W. (2013). You write but others read: Common methodological misunderstandings in PLS and related methods. In H. Abdi, W. W. Chin, V. Esposito Vinzi, G. Russolillo, & L. Trinchera (Eds.), *New perspectives in partial least squares and related methods* (pp. 31–64). New York, NY: Springer.

Martensen, A., & Grønholdt, L. (2010). Measuring and managing brand equity: A study with focus on product and service quality in banking. *International Journal of Quality and Service Sciences, 2,* 300–316.

Martilla, J. A., & James, J. C. (1977). Importance-performance analysis. *Journal of Marketing, 41,* 77–79.

Masyn, K. E. (2013). Latent class analysis ad finite mixture modeling. In P. Nathan & T. Little (Eds.), *The Oxford handbook of quantitative methods* (pp. 551–611). New York, NY: Oxford University Press.

Mateos-Aparicio, G. (2011). Partial least squares (PLS) methods: Origins, evolution, and application to social sciences. *Communications in Statistics—Theory and Methods, 40,* 2305–2317.

Matthews, L., Sarstedt, M., Hair, J. F., & Ringle, C. M. (2016). Identifying and treating unobserved heterogeneity with FIMIX-PLS: Part II—A case study. *European Business Review, 28,* 208–224.

McLachlan, G. J., & Peel, D. (2000). *Finite mixture models.* New York, NY: Wiley.

Meyer, J. P., & Allen, N. J. (1991). A three-component conceptualization of organizational commitment. *Human Resource Management Review, 1,* 61–89.

Michell, J. (2013). Constructs, inferences and mental measurement. *New Ideas in Psychology, 31,* 13–21.

Minkman, E., Rutten, M. M., & van der Sanden, M. C. A. (2017). Acceptance of mobile technology for citizen science in water resource management. *Journal of Irrigation and Drainage Engineering.* Advance online publication. doi:10.1061/(ASCE)IR.1943-4774.0001043

Monecke, A., & Leisch, F. (2012). semPLS: Structural equation modeling using partial least squares. *Journal of Statistical Software, 48.* Retrieved from http://www.jstatsoft.org

Money, K. G., Hillenbrand, C., Henseler, J., & Da Camara, N. (2012). Exploring unanticipated consequences of strategy amongst stakeholder segments: The case of a European revenue service. *Long Range Planning, 45,* 395–423.

Muthén, B. O. (1989). Latent variable modeling in heterogeneous populations. *Psychometrika, 54,* 557–585.

Nitzl, C. (2016). The use of partial least squares structural equation modelling (PLS-SEM) in management accounting research: Directions for future theory development. *Journal of Accounting Literature, 37,* 19–35.

Nunnally, J. C., & Bernstein, I. (1994). *Psychometric theory.* New York, NY: McGraw-Hill.

Pai, H. C. (2016). Development and validation of the simulation learning effectiveness scale for nursing students. *Journal of Clinical Nursing, 25,* 3373–3381.

Pedrosa, R. B. D. S., Rodrigues, R. C. M., Oliveira, H. C., & Alexandre, N. M. C. (2016). Construct validity of the Brazilian version of the self-efficacy for appropriate medication adherence scale. *Journal of Nursing Measurement, 24,* 18E–31E.

Peng, D. X., & Lai, F. (2012). Using partial least squares in operations management research: A practical guideline and summary of past research. *Journal of Operations Management, 30,* 467–480.

Peter, J.-P., & Churchill, G. A. (1986). Relationships among research design choices and psychometric properties of rating scales: A meta analysis. *Journal of Marketing Research, 23,* 1–10.

Petter, S., Straub, D., & Rai, A. (2007). Specifying formative constructs in information systems research. *MIS Quarterly, 31,* 623–656.

Polites, G. L., Roberts, N., & Thatcher, J. B. (2012). Conceptualizing models using multidimensional constructs: A review and guidelines for their use. *European Journal of Information Systems, 21,* 22–48.

Raithel, S., & Schwaiger, M. (2014). The effects of corporate reputation perceptions of the general public on shareholder value. *Strategic Management Journal, 36,* 945–956.

Raithel, S., Wilczynski, P., Schloderer, M. P., & Schwaiger, M. (2010). The value-relevance of corporate reputation during the financial crisis. *Journal of Product and Brand Management, 19,* 389–400.

Ramaswamy, V., DeSarbo, W. S., Reibstein, D. J., & Robinson, W. T. (1993). An empirical pooling approach for estimating marketing mix elasticities with PIMS data. *Marketing Science, 12,* 103–124.

Reinartz, W., Haenlein, M., & Henseler, J. (2009). An empirical comparison of the efficacy of covariance-based and variance-based SEM. *International Journal of Research in Marketing, 26,* 332–344.

Richter, N. F., Sinkovics, R. R., Ringle, C. M., & Schlägel, C. (2016). A critical look at the use of SEM in international business research. *International Marketing Review, 33,* 376–404.

Rigdon, E. E. (2005). Structural equation modeling: Nontraditional alternatives. In B. Everitt & D. Howell (Eds.), *Encyclopedia of statistics in behavioral science* (pp. 1934–1941). New York, NY: Wiley.

Rigdon, E. E. (2012). Rethinking partial least squares path modeling: In praise of simple methods. *Long Range Planning, 45,* 341–358.

Rigdon, E. E. (2014). Rethinking partial least squares path modeling: Breaking chains and forging ahead. *Long Range Planning, 47,* 161–167.

Rigdon, E. E. (2016). Choosing PLS path modeling as analytical method in European management research: A realist perspective. *European Management Journal, 34,* 598–605.

Rigdon, E. E., Becker, J.-M., Rai, A., Ringle, C. M., Diamantopoulos, A., Karahanna, E., . . . Dijkstra, T. (2014). Conflating antecedents and formative indicators: A comment on Aguirre-Urreta and Marakas. *Information Systems Research, 25,* 780–784.

Rigdon, E. E., Preacher, K. J., Lee, N., Howell, R. D., Franke, G. R., & Borsboom, D. (2011). Overcoming measurement dogma: A response to Rossiter. *European Journal of Marketing, 45,* 1589–1600.

Rigdon, E. E., Ringle, C. M., & Sarstedt, M. (2010). Structural modeling of heterogeneous data with partial least squares. In N. K. Malhotra (Ed.), *Review of marketing research* (pp. 255–296). Armonk, NY: M. E. Sharpe.

Rigdon, E. E., Ringle, C. M., Sarstedt, M., & Gudergan, S. P. (2011). Assessing heterogeneity in customer satisfaction studies: Across industry similarities and within industry differences. *Advances in International Marketing, 22,* 169–194.

Ringle, C. M., & Sarstedt, M. (2016). Gain more insight from your PLS-SEM results: The importance-performance map analysis. *Industrial Management & Data Systems, 116,* 1865–1886.

Ringle, C. M., Sarstedt, M., & Mooi, E. A. (2010). Response-based segmentation using finite mixture partial least squares: Theoretical foundations and an application to American customer satisfaction index data. *Annals of Information Systems, 8,* 19–49.

Ringle, C. M., Sarstedt, M., & Schlittgen, R. (2014). Genetic algorithm segmentation in partial least squares structural equation modeling. *OR Spectrum, 36,* 251–276.

Ringle, C. M., Sarstedt, M., Schlittgen, R., & Taylor, C. R. (2013). PLS path modeling and evolutionary segmentation. *Journal of Business Research, 66,* 1318–1324.

Ringle, C. M., Sarstedt, M., & Straub, D. W. (2012). A critical look at the use of PLS-SEM in *MIS Quarterly. MIS Quarterly, 36,* iii–xiv.

Ringle, C. M., Wende, S., & Becker, J.-M. (2015). *SmartPLS 3* [Computer software]. Retrieved from http://www.smartpls.com

Ringle, C. M., Wende, S., & Will, A. (2005). *SmartPLS 2* [Computer software]. Retrieved from http://www.smartpls.com

Robins, J. (2012). Partial least squares (Editorial). *Long Range Planning, 45,* 309–311.

Robins, J. (2014). Partial least squares revisited. *Long Range Planning, 47,* 131.

Roldán, J. L., & Sánchez-Franco, M. J. (2012). Variance-based structural equation modeling: Guidelines for using partial least squares in information systems research. In M. Mora, O. Gelman, A. L. Steenkamp, & M. Raisinghani (Eds.), *Research methodologies, innovations and philosophies in software systems engineering and information systems* (pp. 193–221). Hershey, PA: IGI Global.

Rönkkö, M., & Evermann, J. (2013). A critical examination of common beliefs about partial least squares path modeling. *Organizational Research Methods, 16,* 425–448.

Rönkkö, M., McIntosh, C. N., Antonakis, J., & Edwards, J. R. (2016). Partial least squares path modeling: Time for some serious second thoughts. *Journal of Operations Management, 47–48,* 9–27.

Rossiter, J. R. (2002). The C-OAR-SE procedure for scale development in marketing. *International Journal of Research in Marketing, 19,* 305–335.

Rossiter, J. R. (2011). *Measurement for the social sciences. The C-OAR-SE method and why it must replace psychometrics.* Berlin, Heidelberg: Springer.

Salzberger, T., Sarstedt, M., & Diamantopoulos, A. (2016). Measurement in the social sciences: Where C-OAR-SE delivers and where it does not. *European Journal of Marketing, 50,* 1942–1952.

Sánchez, G., Trinchera, L., & Russolillo, G. (2015). *R Package PLSPM: Tools for Partial Least Squares Path Modeling (PLS-PM)* (version 0.4.7) [Computer software]. Retrieved from http://cran.r-project.org/web/packages/plspm

Sarstedt, M. (2008). A review of recent approaches for capturing heterogeneity in partial least squares path modelling. *Journal of Modelling in Management, 3,* 140–161.

Sarstedt, M., Becker, J.-M., Ringle, C. M., & Schwaiger, M. (2011). Uncovering and treating unobserved heterogeneity with FIMIX-PLS: Which model selection criterion provides an appropriate number of segments? *Schmalenbach Business Review, 63,* 34–62.

Sarstedt, M., Diamantopoulos, A., & Salzberger, T (2016). Should we use single items? Better not. *Journal of Business Research, 69,* 3199–3203.

Sarstedt, M., Diamantopoulos, A., Salzberger, T., & Baumgartner, P. (2016). Selecting single items to measure doubly-concrete constructs: A cautionary tale. *Journal of Business Research, 69,* 3159–3167.

Sarstedt, M., Ringle, C. M., & Hair, J. F. (2017). Partial least squares. In C. Homburg, M. Klarmann, & A. Vomberg (Eds.), *Handbook of marketing research.* Berlin, Heidelberg: Springer.

Sarstedt, M., Hair, J. F., Ringle, C. M., Thiele, K. O., & Gudergan, S. P. (2016). Measurement issues with PLS and CB-SEM: Where the bias lies! *Journal of Business Research, 69,* 3998–4010.

Sarstedt, M., Henseler, J., & Ringle, C. M. (2011). Multi-group analysis in partial least squares (PLS) path modeling: Alternative methods and empirical results. *Advances in International Marketing, 22,* 195–218.

Sarstedt, M., & Mooi, E. A. (2014). *A concise guide to market research. The process, data, and methods using IBM SPSS statistics* (2nd ed.). Berlin, Germany: Springer.

Sarstedt, M., & Ringle, C. M. (2010). Treating unobserved heterogeneity in PLS path modelling: A comparison of FIMIX-PLS with different data analysis strategies. *Journal of Applied Statistics, 37,* 1299–1318.

Sarstedt, M., Ringle, C. M., Henseler, J., & Hair, J. F. (2014). On the emancipation of PLS-SEM. A commentary on Rigdon (2012). *Long Range Planning, 47,* 154–160.

Sarstedt, M., Ringle, C. M., Smith, D., Reams, R., & Hair, J. F. (2014). Partial least squares structural equation modeling (PLS-SEM): A useful tool for family business researchers. *Journal of Family Business Strategy, 5,* 105–115.

Sarstedt, M., & Schloderer, M. P. (2010). Developing a measurement approach for reputation of non-profit organizations. *International Journal of Nonprofit & Voluntary Sector Marketing, 15,* 276–299.

Sarstedt, M., Schwaiger, M., & Ringle, C. M. (2009). Do we fully understand the critical success factors of customer satisfaction with industrial goods? Extending Festge and Schwaiger's model to account for unobserved heterogeneity. *Journal of Business Market Management, 3,* 185–206.

Sarstedt, M., Wilczynski, P., & Melewar, T. (2013). Measuring reputation in global markets: A comparison of reputation measures' convergent and criterion validities. *Journal of World Business, 48,* 329–339.

Satterthwaite, F. E. (1946). An approximate distribution of estimates of variance components. *Biometrics Bulletin, 2,* 110–114.

Schlägel, C., & Sarstedt, M. (2016). Assessing the measurement invariance of the four-dimensional cultural intelligence scale across countries: A composite model approach. *European Management Journal, 34,* 633–649.

Schlittgen, R., Ringle, C. M., Sarstedt, M., & Becker, J.-M. (2016). Segmentation of PLS path models by iterative reweighted regressions. *Journal of Business Research, 69,* 4583–4592.

Schloderer, M. P., Sarstedt, M., & Ringle, C. M. (2014). The relevance of reputation in the nonprofit sector: The moderating effect of socio-demographic characteristics. *International Journal of Nonprofit and Voluntary Sector Marketing, 19,* 110–126.

Schneeweiß, H. (1991). Models with latent variables: LISREL versus PLS. *Statistica Neerlandica, 45,* 145–157.

Schönemann, P. H., & Wang, M.-M. (1972). Some new results on factor indeterminacy. *Psychometrika, 37,* 61–91.

Schwaiger, M. (2004). Components and parameters of corporate reputation: An empirical study. *Schmalenbach Business Review, 56,* 46–71.

Schwaiger, M., Raithel, S., & Schloderer, M. P. (2009). Recognition or rejection: How a company's reputation influences stakeholder behavior. In J. Klewes & R. Wreschniok (Eds.), *Reputation capital: Building and maintaining trust in the 21st century* (pp. 39–55). Berlin, Heidelberg: Springer.

Schwaiger, M., Sarstedt, M., & Taylor, C. R. (2010). Art for the sake of the corporation: Audi, BMW Group, DaimlerChrysler, Montblanc, Siemens, and Volkswagen help explore the effect of sponsorship on corporate reputations. *Journal of Advertising Research, 50,* 77–90.

Schwarz, G. (1978). Estimating the dimensions of a model. *Annals of Statistics, 6,* 461–464.

Shmueli, G. (2010). To explain or to predict? *Statistical Science, 25,* 289–310.

Slack, N. (1994). The importance-performance matrix as a determinant of improvement priority. *International Journal of Operations and Production Management, 44,* 59–75.

Spearman, C. (1927). *The abilities of man.* New York, NY: Macmillan.

Squillacciotti, S. (2005). Prediction-oriented classification in PLS path modeling. In T. Aluja, J. Casanovas, V. Esposito Vinzi, & M. Tenenhaus (Eds.), *PLS and marketing: Proceedings of the 4th International Symposium on PLS and related methods* (pp. 499–506). Paris, France: DECISIA.

Squillacciotti, S. (2010). Prediction-oriented classification in PLS path modeling. In V. Esposito Vinzi, W. W. Chin, J. Henseler, & H. Wang (Eds.), *Handbook of partial least squares: Concepts, methods and applications in marketing and related fields* (pp. 219–233). Berlin, Germany: Springer.

Steenkamp, J. B. E. M., & Baumgartner, H. (1998). Assessing measurement invariance in cross national consumer research. *Journal of Consumer Research, 25,* 78–107.

Steinley, D. (2003). Local optima in k-means clustering: What you don't know may hurt you. *Psychological Methods, 8,* 294–304.

Temme, D., & Diamantopoulos, A. (2016). Higher-order models with reflective indicators: A rejoinder to a recent call for their abandonment. *Journal of Modelling in Management, 11,* 180–188.

Temme, D., Kreis, H., & Hildebrandt, L. (2010). A comparison of current PLS path modeling software: Features, ease-of-use, and performance. In V. Esposito Vinzi, W. W. Chin, J. Henseler, & H. Wang (Eds.), *Handbook of partial least squares: Concepts, methods and applications* (pp. 737–756). Heidelberg, Germany: Springer.

Tenenhaus, M., Esposito Vinzi, V., Chatelin, Y.-M., & Lauro, C. (2005). PLS path modeling. *Computational Statistics & Data Analysis, 48,* 159–205.

Test&Go. (2006). *SPAD-PLS version 6.0.0* [Computer software]. Paris, France.

Thurstone, L. L. (1947). *Multiple factor analysis.* Chicago, IL: University of Chicago Press.

Valencia, J. L., & Diaz-Llanos, F. J. (2003). *Regresión PLS en las ciencias experimentales* [PLS regression in the experimental sciences]. Madrid, Spain: Editorial Complutense.

Vandenberg, R. J., & Lance, C. E. (2000). A review and synthesis of the measurement invariance literature: Suggestions, practices, and recommendations for organizational research. *Organizational Research Methods, 3,* 4–70.

Völckner, F., Sattler, H., Hennig-Thurau, T., & Ringle, C. M. (2010). The role of parent brand quality for service brand extension success. *Journal of Service Research, 13,* 359–361.

Wedel, M., & Kamakura, W. (2000). *Market segmentation. Conceptual and methodological foundations* (2nd ed.). Norwell, MA: Kluwer Academic.

Welch, B. L. (1947). The generalization of "student's" problem when several different population variances are involved. *Biometrika, 34,* 28–35.

Wetzels, M., Odekerken-Schroder, G., & van Oppen, C. (2009). Using PLS path modeling for assessing hierarchical construct models: Guidelines and empirical illustration. *MIS Quarterly, 33,* 177–195.

Wickens, M. R. (1972). A note on the use of proxy variables. *Econometrica, 40,* 759–761.

Wilden, R., & Gudergan, S. (2015). The impact of dynamic capabilities on operational marketing and technological capabilities: Investigating the role of environmental turbulence. *Journal of the Academy of Marketing Science, 43,* 181–199.

Willaby, H. W., Costa, D. S. J., Burns, B. D., MacCann, C., & Roberts, R. D. (2015). Testing complex models with small sample sizes: A historical overview and empirical demonstration of what partial least squares (PLS) can offer differential psychology. *Personality and Individual Differences, 84,* 73–78.

Wold, H. O. A. (1966). Estimation of principal components and related methods by iterative least squares. In P. R. Krishnaiah (Ed.), *Multivariate analysis* (pp. 391–420). New York, NY: Academic Press.

Wold, H. O. A. (1973). Nonlinear iterative partial least squares (NIPALS) modeling: Some current developments. In P. R. Krishnaiah (Ed.), *Multivariate analysis III* (pp. 383–407). New York, NY: Academic Press.

Wold, H. O. A. (1980). Model construction and evaluation when theoretical knowledge is scarce: Theory and application of partial least squares. In J. Kmenta & J. B. Ramsey (Eds.), *Evaluation of econometric models* (pp. 47–74). New York, NY: Academic Press.

Wold, H. O. A. (1982). Soft modeling: The basic design and some extensions. In K. G. Jöreskog & H. Wold (Eds.), *Systems under indirect observations: Part II* (pp. 1–54). Amsterdam, Netherlands: North-Holland.

Wold, H. O. A. (1985). Partial least squares. In S. Kotz & N. L. Johnson (Eds.), *Encyclopedia of statistical sciences* (pp. 581–591). New York, NY: John Wiley.

Zhang, Y., & Schwaiger, M. (2009). An empirical research of corporate reputation in China. *Communicative Business, 1,* 80–104.

Author Index

Subject Index

Exhibits are indicated by *italicized* page numbers